# Indwelling and Implantable Pressure Transducers*

Editors:

**David G. Fleming**
Professor of Biomedical Engineering
Case Western Reserve University
Cleveland, Ohio

**Wen H. Ko**
Director, Engineering Design Center
Professor of Biomedical Engineering
Case Western Reserve University
Cleveland, Ohio

**Michael R. Neuman**
Associate Professor of Biomedical Engineering
in Reproductive Biology
Case Western Reserve University
Cleveland, Ohio

*Based on workshop held in Cleveland, Ohio on December 4 and 5, 1975, sponsored by the Biotechnology Resources Branch (RR-00857) and the National Institute of General Medical Sciences (GM-14267) of the National Institutes of Health.

CRC Press
Taylor & Francis Group
Boca Raton London New York

CRC Press is an imprint of the
Taylor & Francis Group, an informa business

First published 1977 by CRC Press
Taylor & Francis Group
6000 Broken Sound Parkway NW, Suite 300
Boca Raton, FL 33487-2742

Reissued 2018 by CRC Press



**Library of Congress Cataloging in Publication Data**
Main entry under title:

Indwelling and implantable pressure transducers.

Bibliography: p.
Includes indexes.
I. Pressure transducers, Biomedical – Congresses.
I. Fleming, David G., 1926– II. Ko, Wen H.,
1923– III. Neuman, Michael R. IV. United States.
National Institutes of Health. Biotechnology Resources
Branch. V. United States. National Institute of General
Medical Sciences. [DNLM: 1. Transducers – Congresses.
2 Biomedical engineering – Congresses. QT34 I42 1975]
R857.T7152 610'.28 76-48168
ISBN-0-8493-5195-2

A Library of Congress record exists under LC control number: 76048168



ISBN 13: 978-1-315-89441-6 (hbk)
ISBN 13: 978-1-351-07351-6 (ebk)

Visit the Taylor & Francis Web site at http://www.taylorandfrancis.com and the
CRC Press Web site at http://www.crcpress.com

# PREFACE

The information presented in this publication is based on a 2-day workshop held at Case Western Reserve University on December 4 and 5, 1975. We have attempted to retain the flavor of the sessions while giving each author the opportunity to modify and update his original contribution. Almost all of the informal panel discussions have been deleted and/or incorporated into the text in other ways.

The workshop was organized by the senior members of the Biomedical Electronics Resource, with the generous assistance of an external program committee consisting of the following individuals:

William R. Baker, Jr. – BRB, National Institutes of Health

Dean Franklin – University of California at San Diego

Alan Kahn – Medtronics, Incorporated

Harold Sandler – NASA

Louis Statham – Founder of Statham Instruments Division of Gould Corporation

Consultants:

Bernard Schacter – NIGMS, National Institutes of Health

Ronald Chez – NICHD, National Institutes of Health

A prospectus was submitted to and subsequently funded by the Biotechnology Resources Branch, Division of Research Resources, National Institutes of Health. This generous support provided funding for speakers and other conference expenses. The entire workshop was initially transcribed, and typed copies of the presentations were sent to the contributors within weeks of the last session. Arrangements were then made with CRC Press to provide rapid manuscript processing and final production. Through the cooperation of the authors, editors, and production personnel, this book appears prior to the first anniversary of the workshop.

Indwelling and implantable devices represent examples of high technology applied to the basic health sciences and clinical medicine. While basic principles are primarily invariant, the ways they are applied may change rapidly and often render obsolete both the devices and the data related to their use. Pressure transducer technology has made significant progress over the past decade to the point that the idea of constructing a device which could be placed in any hollow viscous of a living organism to register and transmit a signal related to the pressure in that structure is no longer considered beyond the limits of existing technology.

The implications and potential applications of miniature pressure sensors constitute much of what follows in this volume. A discussion of the problems associated with achieving this technology is also included.

No undertaking of this magnitude can be completed without the cooperation of many individuals, both inside and outside the Biomedical Electronics Resource.

Technical assistance in organizing the workshop was provided by the Verve Research Corporation. Local arrangements were managed by Mr. Sidney Williams and the staff of the Alumni Relations Office of Special Events, who made insurmountable problems merely difficult and reduced difficult problems to the nonexistent.

We wish to express our appreciation to the staff of the Engineering Design Center, including Dr. James Topich, Scott Boettcher, John Homa, C. W. Poon, Eddie Cheng, and Spiro Vamvakas, who set up the displays, acted as hosts and tour guides, and provided reassurance to the secretaries and typists who were attempting to convert the jargon of the speakers to a written text.

Special thanks is reserved for Ms. Lois Schweitzer and, in particular, Ms. Margaret Butler who collected, sorted and typed the assorted documents associated with the workshop.

A final note of gratitude to Ms. Denise Lee of CRC Press for her help and patient understanding in editing the manuscripts.

To complete our acknowledgments, we would like to cite the after-dinner speech of Dean William Drucker of the University of Virginia School of Medicine, who held the participants spellbound with his highly opinionated view of the origins of Biomedical Engineering.

This volume is dedicated to our families, bless them all.

D. G. Fleming
Wen H. Ko
Michael R. Neuman
August 1976

## THE EDITORS

**David G. Fleming,** M.D., Ph.D., is Professor of Biomedical Engineering in Pediatrics and Biomedical Engineering. He is also the Associate Director of the Biomedical Electronics Resource, with responsibility for educational programs.

Dr. Fleming received his A.B. in zoology in 1948 and his Ph.D. in physiology in 1952, both from the University of California at Berkeley. He received his M.D. from Case Western Reserve University in 1973 and was a Pediatric Resident through 1975.

From 1968 to 1970, Dr. Fleming was President of the IEEE Group on Engineering in Medicine and Biology. During the same time, Dr. Fleming also served as Chairman of the ASEE's Bio-and-Medical Engineering Committee. He also was a member of the Joint Committee on Engineering in Medicine and Biology from 1965 to 1970, acting as chairman in 1967.

Dr. Fleming's major research interests are biomedical instrumentation for high risk infants and pediatric pulmonary medicine. He is currently a Clinical Fellow in Pulmonary Pediatrics. He has authored many articles in these areas and holds several patents in biomedical instrumentation.

**Wen H. Ko**, Ph.D., is Director of the Engineering Design Center and Professor of Electrical and Biomedical Engineering at Case Western Reserve University, Cleveland, Ohio.

Dr. Ko graduated from the National Amoy University, Fukien, China, in 1946 with a B.S. degree in electrical engineering. He received his M.S. and Ph.D. degrees in electrical engineering in 1956 and 1959, respectively, from Case Institute of Technology, Cleveland, Ohio.

Dr. Ko, who is a senior member of the Institute of Electrical and Electronics Engineers, has served as a consultant to the Research Grants Branch of the National Institutes of Health and various industries, including North American Manufacturing Company.

Dr. Ko holds several patents in the area of solid state electronics. His major research interest involves biomedical instrumentation in the areas of implant telemetry and stimulation systems, and patient monitoring systems.

**Michael R. Neuman**, M.D., Ph.D., is an Associate Professor of Biomedical Engineering in Reproductive Biology and holds a joint appointment in Biomedical Engineering at Case Western Reserve University, Cleveland, Ohio. He is also a member of the Engineering Design Center at Case Western Reserve University.

Dr. Neuman received his B.S., M.S., and Ph.D. degrees in electrical engineering from Case Institute of Technology in 1961, 1963, and 1966, respectively. He received an M.D. degree from the Case Western Reserve University School of Medicine in 1974.

Dr. Neuman is a research associate in the Departments of Obstetrics and Gynecology at University Hospitals of Cleveland and Cleveland Metropolitan General Hospital. He is a member of numerous professional societies, including the Society for Gynecologic Investigation and the Association for the Advancement of Medical Instrumentation.

Dr. Neuman's research interests have been in the area of biomedical instrumentation as applied to the reproductive systems in animal models and man. He has published over 30 papers in the areas of biomedical engineering and obstetrics and gynecology, and holds one patent.

## CONTRIBUTORS

**James B. Angell, Sc.D.**
Professor and Associate Chairman
Electrical Engineering Department
Stanford University
Stanford, California 94305

**Hank Baig, M.S.E.E.**
Research Engineer
Peter Bent Brigham Hospital
New England Regional Primate Research Center
1 Pine Hill Drive
Southboro, Massachusetts 01772

**Scott Boettcher, B.S.E.E.**
Staff Engineer
Engineering Design Center
Case Western Reserve University
Cleveland, Ohio 44106

**Gordon Borkat, M.D.**
Director, Pediatric Cardiac Catherization Laboratory
University Hospitals
2065 Adelbert Road
Cleveland, Ohio 44106

**Jess W. Burns, B.S.**
Product Manager
Medical Pressure Transducers
Gould Corporation
Statham Instruments Division
2230 Statham Boulevard
Oxnard, California 93030

**Eddie M. Cheng, M.S.**
Research Assistant
Engineering Design Center
Case Western Reserve University
Cleveland, Ohio 44106

**John Chubbuck, M.S.**
Principal Professional Staff
Applied Physics Laboratory
Johns Hopkins University
Johns Hopkins Road
Laurel, Maryland 20810

**Alastair Connell, M.D.**
Professor of Physiology
Director, Division of Digestive Diseases
University of Cincinnati
Cincinnati, Ohio 45267

**Thomas B. Fryer, Ph.D.**
Electronic Instrument Development Branch
National Aeronautics and Space Administration
Ames Research Center
Moffett Field, California 94035

**Leslie A. Geddes, M.E., Ph.D.**
Director, Biomedical Engineering Center
Showalter Distinguished Professor
Purdue University
West Lafayette, Indiana 47907

**Jaroslav Hynecek, Ph.D.**
Chief Engineer
Engineering Design Center
Case Western Reserve University
Cleveland, Ohio 44106
presently with
Central Research Laboratories of Texas Instruments
Dallas, Texas 75222

**Gordon Jacobs, M.S.E.E.**
Artificial Heart Project Leader
Department of Artificial Organs
Division of Research
Cleveland Clinic Foundation
9500 Euclid Avenue
Cleveland, Ohio 44106

**Raymond J. Kiraly, M.S.M.E.**
Director of Engineering
Department of Artificial Organs
Division of Research
Cleveland Clinic Foundation
9500 Euclid Avenue
Cleveland, Ohio 44106

**Paul S. Lederer, B.E.E.**
Chief, Components and Applications Section
Electronic Technology Division
National Bureau of Standards
Washington, D.C. 20234

**Ronald J. Lorig, Ph.D.**
Assistant Professor of Neurosurgery-Biomedical Engineering
Case Western Reserve University
Cleveland, Ohio 44106

**Ernest P. McCutcheon, M.D.**
Director, Cardiovascular Research Laboratory
Biomedical Research Division
National Aeronautics and Space Administration
Ames Research Center
Moffett Field, California 94035

**Richard S. Muller, Ph.D.**
Professor, Department of Electrical Engineering and Computer Sciences
University of California
Berkeley, California 94720

**Yukihiko Nosé, M.D., Ph.D.**
Head, Department of Artificial Organs
Division of Research
Cleveland Clinic Foundation
9500 Euclid Avenue
Cleveland, Ohio 44106

**Timothy A. Nunn, M.S.**
Research Assistant and Ph.D. Candidate
Electrical Engineering Department
Stanford University
Stanford, California 94305

**Thomas Patrick, B.S.E.E.**
Senior Research Engineer
Peter Bent Brigham Hospital
New England Regional Primate Research Center
1 Pine Hill Drive
Southboro, Massachusetts 01772

**Erik L. Ritman, M.D., Ph.D.**
Consultant, Cardiovascular Research Unit
Department of Cardiovascular Diseases and Internal Medicine
Biophysical Sciences Unit
Department of Physiology and Biophysics
Mayo Clinic/Mayo Foundation
Assistant Professor of Physiology
Mayo Medical School
Rochester, Minnesota 55901

**Harold Sandler, M.D.**
Chief, Biomedical Research Division
National Aeronautics and Space Administration
Washington, D.C. 20546

**H. L. Stone, Ph.D.**
Professor of Physiology and Biophysics
Chief, Cardiovascular Control Section
Division of Comparative Marine Neurobiology
The Marine Biomedical Institute
University of Texas Medical Branch
Galveston, Texas 77550

**Gerald W. Timm, Ph.D.**
Associate Professor, Department of Neurology
University of Minnesota Hospitals
Minneapolis, Minnesota 55455
presently
Vice President, Research and Engineering
American Medical Systems, Inc.
3312 Gorham Avenue
Minneapolis, Minnesota 55426

**Stephen F. Vatner, M.D.**
Associate Professor of Medicine
Harvard Medical School
Cambridge, Massachusetts 02138
Peter Bent Brigham Hospital
New England Regional Primate Research Center
1 Pine Hill Drive
Southboro, Massachusetts 01772

**Lawrence J. Viernstein, Ph.D.**
Principal Professional Staff
Applied Physics Laboratory
Johns Hopkins University
Johns Hopkins Road
Laurel, Maryland 20810

**A. Earl Walker, M.D.**
Professor of Surgery
University of New Mexico School of Medicine
Albuquerque, New Mexico 87131
Professor Emeritus of Neurological Surgery
Johns Hopkins University
Baltimore, Maryland 21205

**Earl H. Wood, M.D., Ph.D.**
Chairman, Biophysical Sciences Unit
Department of Physiology and Biophysics
Mayo Clinic/Mayo Foundation
Professor of Physiology and Medicine
Mayo Medical School
Rochester, Minnesota 55901

**Keming W. Yeh, Ph.D.**
Microelectronics Center
Xerox Corporation
2383 Utah Avenue
El Segundo, California 90245

# TABLE OF CONTENTS

## NEEDS AND POTENTIAL APPLICATIONS OF INDWELLING PRESSURE TRANSDUCERS



## PRESENT APPLICATIONS OF INDWELLING PRESSURE TRANSDUCERS

# *Needs and Potential Applications of Indwelling Pressure Transducers*

# INTRODUCTION

**David G. Fleming**

This book and the workshop which led to its preparation grew out of the personal needs of the editors to obtain a reasonably accurate estimate of the current level of development of pressure transducers designed for use in situations requiring the measurement of pressure in living organisms. Most of the presentations in this volume deal with devices designed for implantable or (at least) indwelling applications.

The last 15 years have seen an explosion in the high technology associated with the development of solid state devices. Many of these advances have led to the development of specific instrumentation for the biomedical sciences. Unfortunately, in most areas the failure rate has exceeded the success rate and has been a source of concern to designer and user alike. Most observers consider that the primary cause of this less than perfect record is the significant and often unbridged gap between the engineer-designer on one side and the scientist-clinician on the other.

Few people, including many professionals in the university and governmental communities, fully appreciate the complexity of the problems associated with the conversion of an instrument which works well in the hands of the originator under restricted laboratory conditions to a rugged, safe, commercially produced, reliable device which still continues to function in the hands of the eager, but unknowing, user. There have been several recent programs devoted to this issue of enhanced technological utilization in general, and these discussions fall beyond the scope of the present volume. On the other hand, the specific problems associated with the design, development, and application of devices for the long-term measurement of intraluminal pressure in research and clinical situations are topics of immediate concern to the editors of and contributors to this volume.

Researchers at the Engineering Design Center, Case Western Reserve University, have been deeply involved in the design, development, and evaluation of implantable, miniature, solid state pressure transducers for the past several years and have acquired considerable first-hand knowledge of the joys and frustrations associated with this process. The concept of bringing designers, developers, manufacturers, and users together to discuss problems of mutual concern quickly matured into a workshop whose purpose is stated below:

> The workshop has as its primary goal the bringing together of designers, builders and users of indwelling pressure transducers, so as to evaluate their present status.
>
> The formal presentations and panels will discuss transducer characteristics, limitations of present devices and suggest new approaches to specific problems in biomedical research. Experiences, technology and fabrication processes will be exchanged, and new approaches for transducer design will be explored. Presentations in the sessions will survey current applications and potential uses of pressure sensors in the life sciences and medicine, covering performance, problems and greater needs. Recommendations will also be made for new transducer designs, in order to improve present performance and open the way for new uses.
>
> The proceedings of the workshop will be published within several months of completion of the meeting.

A program committee was organized and convened by conference call. Potential speakers and panelists were suggested and subsequently contacted.

The reception was favorable, except from commercial manufacturers who declined or were unable to participate. Notable exceptions were Eph Konigsberg whose comments during the actual workshop were stimulating and thought provoking. Amnon Brosh from the Kulite Corporation presented his view of the government/industry/health care system relationship. His remarks were representative of the nonmedical transducer manufacturers. Neither the Konigsberg or Brosh presentations are included in the text, since both felt that their remarks were too informal to be included in the final volume. Louis Statham was taken ill shortly before the sessions were scheduled to begin and was forced to withdraw.

The book is divided into five sections. The initial section is based on a panel chaired by Dr. Alan Kahn and includes this introduction followed by a review of the fundamentals of blood pressure transducers, zero drift, and calibration problems, and concludes with two examples of the uses of implant units in chronic animal studies.

The second section begins with a survey of

present applications of indwelling devices. Specific examples from gastrointestinal motility studies, artificial organ research, neurosurgery, obstetrics, pediatrics, and urology are presented in turn.

A reasonably current list of transducers, suitable for biomedical applications, was added to an abbreviated section on industrial considerations.

The fourth area covered in the volume is related to transducer design and development. It begins with a review of transduction principles, followed by examples of newly developed devices based on these concepts.

The last section of the book evolved from a panel chaired by Dr. William R. Baker, Jr. Dr. Harold Sandler organized his remarks into an explanation of future needs based on a user's point of view.

One of the secondary objects of this introduction is to distinguish indwelling from implantable pressure transducers. While both indwelling and implantable devices are alike in that the two are placed within the animal under study, they differ in that an indwelling device requires a physical channel for the transfer of power-in or signal-out. In some examples, the connection is percutaneous as with intracranial or intravascular units. Other units for gastrointestinal studies or urogenital applications may emerge from an existing stoma, but often require long cables or catheters from the external surface to a target structure. Implantable units, on the other hand, require no physical or hard wire links to the external environment. They are either battery or radiofrequency powered and are coupled to the external environment via a telemetry link.

Most indwelling pressure transducers are designed with a small channel for static calibrations. In contrast, implantable units cannot be calibrated or corrected for zero drift once installed, and the user must rely on the inherent performance characteristics or use an independent system for periodic checks. However, the advantages of implantable units are clear-cut in certain situations. For example, these units provide complete freedom from clutter because they require no wires, catheters, or external plumbing in general. The elimination of the fluid channel reduces the possible chance of hemorrhage to a minimum, and the absence of a percutaneous connection markedly decreases the chance of obtaining infection. Data from free-roaming animals or chronically ill patients under reasonable physiological conditions becomes feasible. Problems associated with thermal drift or reference level changes are also decreased.

There are penalties and problems associated with internal transducer applications. The units must be reliable; electrical and mechanical failure rate has to approach the zero defect level. Should the device fail, it must represent neither a physical nor electrical hazard to the patient. Performance characteristics must exceed any unit currently available with zero drift reduced to 1 mm/month. Linearity should be within 1% and thermal drift reduced to 1 mm/°C. Sensitivity ranges should include both arterial and venous pressures. One or the other range spans the pressures found in other body compartments.

Packaging is important, as the internal environment is hostile to electromechanical devices and usually the converse is true. Tissue reaction (foreign body reaction) must be kept to an absolute minimum, including the formation of thromboemboli in the bloodstream.

The devices should be mechanically rugged, resistant to failure from accidental over-pressure, and small enough to be used in neonatal pediatrics and animal studies. Many of our highest risk premature infants weigh less than 1 kg. New devices should at least be as small as existing catheter-tip transducers which are available in a number four French catheter (see chapter by G. Borkat).

Finally, the cost for an individual unit should be sufficiently low to make it a potentially one-use, disposable item.

An ideal unit would be small, safe, stable, reliable, nonhostile, and inexpensive. It should be easy to implant and capable of remaining in the desired location for an indefinite period of time. How far we have come and how much further there is to go constitute much of what follows.

# FUNDAMENTALS OF BLOOD PRESSURE TRANSDUCERS

**L. A. Geddes**

Blood pressure transducers operate via the electrical detection of the deflection of an elastic member. The fidelity of detection depends on the relationship between the mass, elasticity, and viscous damping present. This chapter describes the requirements for this relationship and the consequences of a mismatch among the relationships between these three quantities.

It is too seldom realized that all pressure transducers operate on the same basic principle. Thus, the following statement is rather concentrated and defines the operation of all pressure transducers. "Blood pressure transducers operate on the basis of the electrical detection of the deflection of an elastic member exposed to pressure." Whether they be catheter-tip, catheter-type, or implanted, they all operate according to the conditions imposed by the second-order differential equation that describes their motion. Among the three types, the only difference is that the three constants in the differential equation have slightly different magnitudes. The constants are mass (M), elasticity (K), and frictional or viscous damping (R). In the implanted and catheter-tip units the viscous forces (R) are very small, thereby permitting a short response time for a sudden change in pressure. Figure 1 illustrates the basic components of a pressure transducer designed for use with a catheter or needle.

As a reminder that catheter-tip pressure transducers are not new, Figure 2 is presented. This device, measuring 3 mm in diameter, is an electrolytic strain gauge in which pressure applied to an elastic membrane moves one electrode closer to the other in the intervening electrolyte. The device was developed by Grünbaum[2] in 1898 and was used to measure the pressure in the right ventricle of the rabbit.

The differential equation describing the operation of all blood pressure transducers, shown in Figure 1, was solved by Otto Frank at the turn of this century. Although high-fidelity mechano-optical systems have been available since then, electronic transducers displaced them in the mid-1940s. Earl Wood of the Mayo Clinic in Rochester, Minnesota was instrumental in gaining acceptance for these devices.

In all pressure transducers the speed of response is determined by the stiffness of the elastic member and the damping and mass of the moving system, which includes the diaphragm, the transducing system connected to it, and the fluid column in front of it. The stiffer the diaphragm, the shorter the response time and the less the electrical output per millimeter of mercury. Stiffness in conventional catheter-type transducers is defined in terms of the volume displacement, which is the number of cubic millimeters of fluid entering the transducer for an applied pressure of 100 mmHg. For a given transducer, choice of the length and diameter of needle or catheter determines the viscous damping. Proper choice of these dimensions can provide the system with a uniform sinusoidal frequency response and minimal overshoot in response to the application of a step function of pressure.[1] Excessive viscous damping, by the use of too narrow or too long a catheter, prolongs the risk time. Figure 3 illustrates the relationship between rise time, damping, overshoot, and sine-wave response.

The desideratum in obtaining high-fidelity recordings with catheter-type transducers is the use of a unit with a low volume displacement, which means that the diaphragm is stiff and hence the natural resonant frequency is high. Choice of the appropriate dimensions for the hydraulic system that couples the transducer to the site of pressure measurement provides the desired damping. Incorrect selection of the dimensions of the coupling system will result in excessively high or low viscous damping. In the former case, the rise time is prolonged and the rapidly changing portions of a pressure waveform are distorted. In the latter case, overshoot and resonant afterwaves will be present in the record.

Regarding the overall system that is to be used for recording pressure, there are three criteria which must be satisfied so that the event being

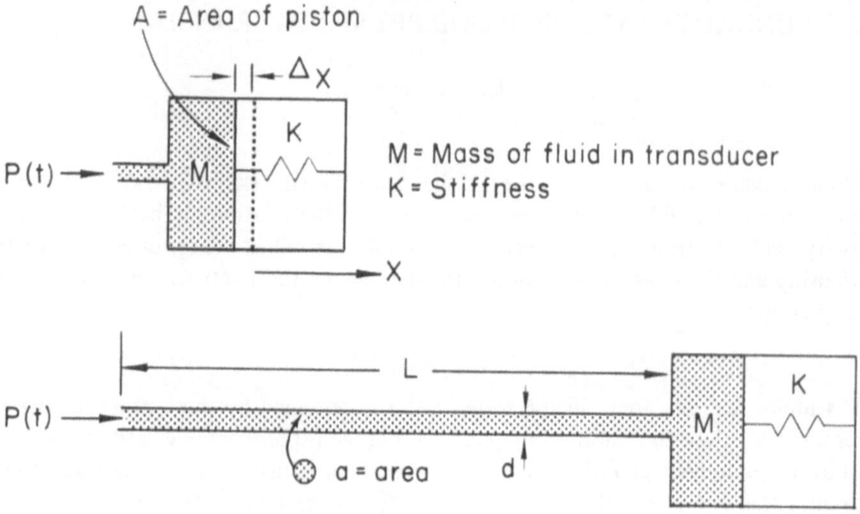


| MOVING MASS | | VISCOUS DRAG | | STIFFNESS | | APPLIED FORCE |
|---|---|---|---|---|---|---|
| $M\frac{d^2x}{dt^2}$ | + | $R\frac{dx}{dt}$ | + | $Kx$ | = | $P(t)$ |

RESONANT FREQUENCY VARIES AS $\sqrt{\frac{\text{STIFFNESS}}{\text{MASS}}}$

STIFFNESS DESCRIBED BY VOLUME DISPLACEMENT ( $V_d$ )

$$V_d = \frac{A\,\Delta x}{\Delta P} = \text{CUBIC MM ENTERING / 100 MM Hg}$$

$$\text{RESONANT FREQUENCY } f_r = \frac{1.4 \times 10^3\, d}{\sqrt{V_d\, L\, \rho}}$$

$$\text{DAMPING } D = \frac{16\pi}{d^3}\sqrt{\frac{3\, L\, V_d}{\eta\, \rho\; 1.33 \times 10^8}}$$

$\eta$ = VISCOSITY $\rho$ = DENSITY

FIGURE 1. Basic principle of all blood pressure transducers. The pressure P (t) is communicated to an elastic member, idealized here by a piston of area (A) pressing against a spring having a stiffness (K). The moving mass (M) includes that of the fluid and the elastic member itself. The stiffness of a transducer is expressed in terms of its volume displacement ($V_d$), being defined as the number of cubic millimeters of fluid entering for the application of a pressure of 100 mmHg. Because of the presence of mass and stiffness, the moving system has a resonant frequency ($f_r$) which, in the absence of damping (R), varies as the ratio of the stiffness to the mass. The presence of viscous drag (R) results in damping which affects the manner by which the transducer responds to sinusoidal and step changes in pressure.

transduced will be reproduced faithfully. The first of the three criteria is amplitude linearity. In other words, if the input signal is doubled the output indication must be doubled. If it is halved or reversed, then the output must respond accordingly. Amplitude linearity is not very difficult to achieve.

The second criterion is a uniform sine-wave frequency response. In other words, it is necessary to provide a sinusoidal frequency response that is uniform out to the highest Fourier frequency component that is considered important for faithful reproduction of the pressure wave. In the case of a system for measuring arterial pressure, six to ten times the cardiac frequency would be adequate. Figure 4 presents information on this

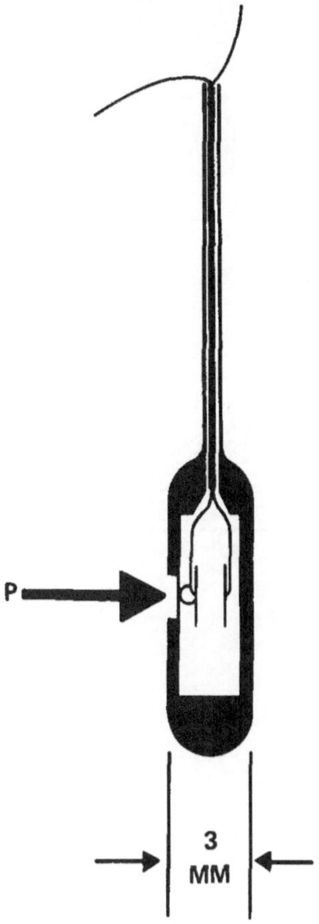


FIGURE 2. The first electrical catheter-tip pressure transducer described by Grunbaum.[2] This transducer consisted of a small chamber of electrolyte between two electrodes; one electrode was mounted on the side of the chamber; the other was mounted on a rubber membrane stretched across a side window. The applied pressure (P) reduced the inter-electrode distance and decreased the resistance measured between the electrodes.

point. Excessive bandwidth does not increase fidelity, but does add considerable cost. It is important to note that the requirement for uniform sinusoidal frequency response relates to that frequency region containing up to about the tenth harmonic of the cardiac frequency. A rise or fall in the sine-wave frequency response curve well beyond this frequency will hardly affect the fidelity of the reproduced wave.

The third criterion states that all of the Fourier components must be delayed by the same amount of time so that they will be added up in the output as they were presented to the input. This requirement can be restated as a need for linear phase shift. The importance of phase shift means that if this criterion is not satisfied, notches in the pressure wave, such as the dicrotic notch and the anacrotic notch, may not be in their proper temporal position in the reproduced wave. In general, if a system has a uniform sine-wave response, phase distortion is usually negligible.

As stated earlier, catheter-tip and implanted transducers, in which the elastic member is in direct contact with the bloodstream, are immune from the high viscous damping that exists in catheter and needle systems. In addition, their miniature size, coupled with low damping, result in the availability of a wide sine-wave frequency response, offering the opportunity for the maximum fidelity.

As a potential user of an implantable pressure transducer, there are certain operational characteristics that a physiologist must know. For example pressure, like voltage, is always measured with respect to some reference. If given an implantable pressure transducer, a physiologist would need to know the identity of the reference or try to determine it if accessible. The reference could be some internal organ or body cavity. In this way differential pressure could be obtained. If differential pressure could not be measured, the researcher would probably determine if the reference pressure had been made equal to ambient pressure at the time of manufacture. The physiologist may also want to know at which temperature the system was sealed, because the transducer will have both temperature and pressure errors if it is moved from these sealing points.

Calibration and long-term stability are important operational characteristics, and high performance in these areas is expensive. It is therefore necessary to know about zero drift and possible calibration sensitivity changes. Zero drift or sensitivity change is not a major concern if the transducer can be recalibrated from outside of the body at any time. Most who conduct physiological studies like to standardize their equipment before the study; it is therefore unimportant if there have been a few millimeters of mercury or more drift over a period of weeks, providing the zero point and pressure sensitivity can be checked prior to, during, or after an experiment.

The present author is interested in the susceptibility of an implanted transducer to externally applied currents. For example, when studies are conducted on ventricular defibrillation, a considerable current is sent through the heart. An implanted transducer (or its wires) may be in the current field and it would be desirable to know if the device will be destroyed during defibrillation.

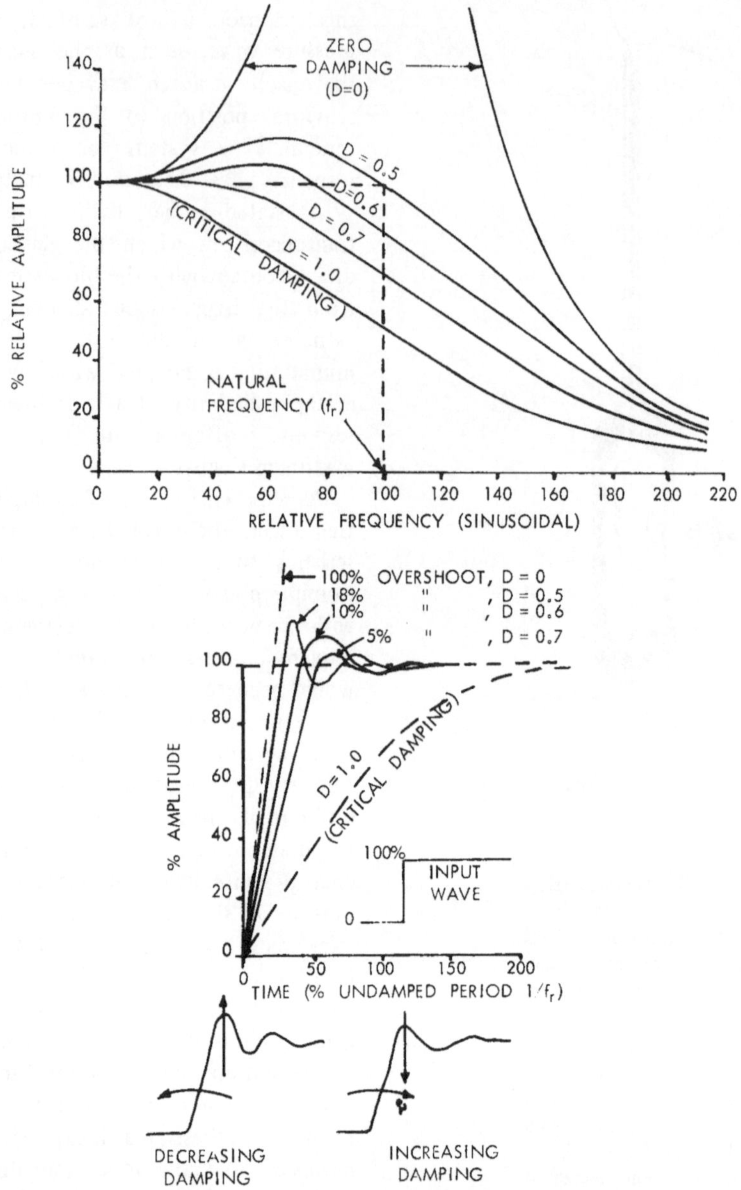


FIGURE 3. The sinusoidal hydraulic frequency response of a pressure transducer with different degrees of damping from zero to critical is shown in the upper diagram. With zero damping, the amplitude of motion of the elastic member would be infinite if the frequency of the applied pressure is equal to the natural resonant frequency ($f_r$). Increasing damping provides a more uniform sinusoidal frequency response. In the lower diagram is shown the response to an input pressure wave which instantaneously rises from zero to its final value (100%). The deflection (% amplitude) of the elastic member is shown for various degrees of damping. Note that for critical damping (the least damping for zero overshoot), the response time is prolonged. Note also that with zero damping, the response time is determined by the resonant frequency, but the overshoot is 100%. Increasing the damping diminishes the overshoot and prolongs the rise time as shown.

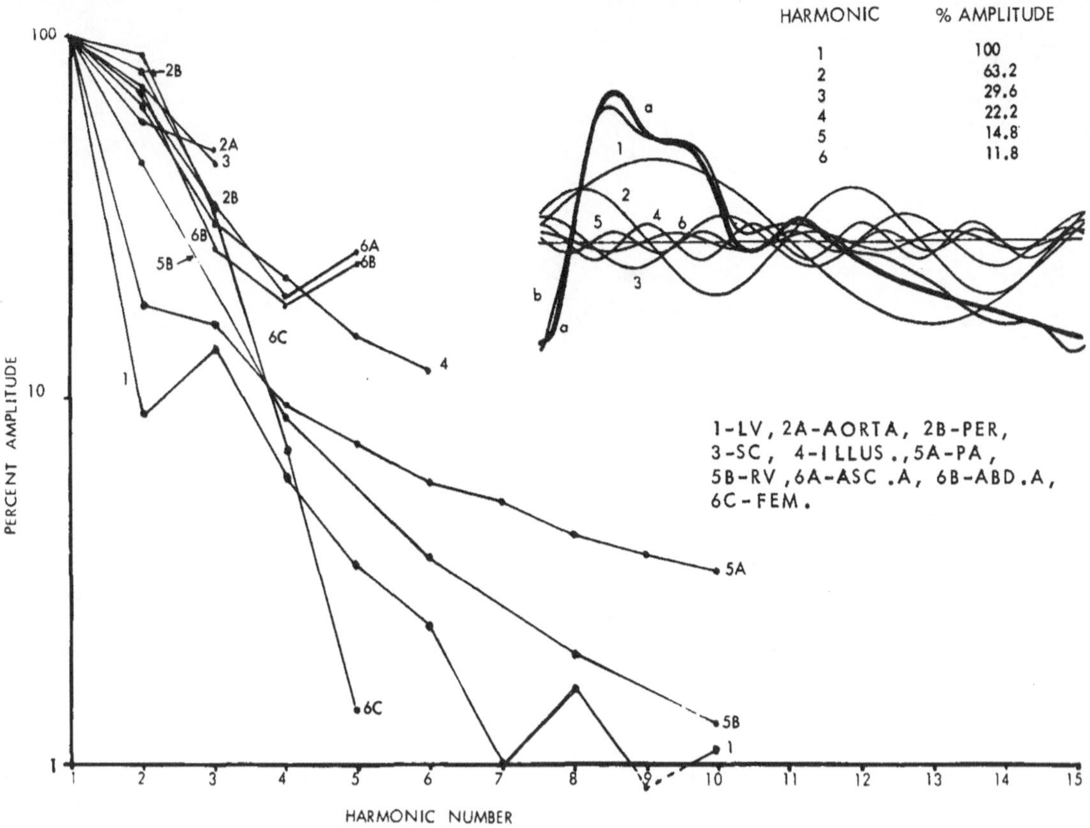


FIGURE 4. Fourier analysis of arterial pressure waves. The inset shows reconstruction of the arterial pressure wave (a) by summing the first six harmonics of the Fourier spectrum. Note that the amplitudes of the higher harmonics diminish with increasing harmonic number. The graph illustrates the harmonic content of various pressure waves: 1 = left ventricle, 2A = aorta, 2B = peripheral pulse, 3 = subclavian, 4 = inset, 5A = pulmonary artery, 5B = right ventricle, 6A = ascending aorta, 6B = abdominal aorta, 6C = femoral artery. Note that, in all cases, the amplitudes of the harmonics diminish with increasing harmonic number.

Many researchers apply current to measure physiological events by impedance, and it is important to know if the operation of the transducer is going to be altered by externally applied audiofrequency current, and if so, in what way.

Because the elastic member deflected by pressure is in contact with blood, it would also be valuable to know if the diaphragm is connected to the electronics or if the metal case can be used as an electrode. Sometimes this feature is very useful when recording or telemetering bioelectric events.

Implantable pressure transducers are going to be expensive for some time; therefore, knowledge of the expected lifetime of the device would be valuable. Such information is not only of economic importance, but it provides data allowing the design of experiments. Since surgical implantation time and animals are both expensive, knowledge of transducer lifetime allows optimum planning of an experiment.

The author's particular use for an implanted pressure transducer can be described in order of increasing difficulty of use. Such a transducer would first be used for measurement of aortic pressure, then left and right ventricular pressures, and then venous pressure. Measurements would be made at 37°C to as low as 20°C. Drift is not a primary concern as long as the location of zero pressure and the output for a known pressure can be identified.

## REFERENCES


1. **Geddes, L. A.,** *The Direct and Indirect Measurement of Blood Pressure,* Year Book Medical, Chicago, 1970.
2. **Grünbaum, O. F.,** On a new method of recording alterations of pressure, *J. Physiol.,* 22, XLIX-LI, 1897.

# IMPLANTABLE PRESSURE TRANSDUCERS – FUTURE NEEDS AND APPLICATIONS OF TOTALLY IMPLANTABLE UNITS

**H. L. Stone**

The cardiovascular use of implantable pressure transducers has been primarily directed toward the arterial system. The reason for this has been the lack of zero stability with existing transducers. At the current time implantation studies have ranged from the chimpanzee to the small *Macaca mulatta.* The zero drift of the pressure transducers increases with decreasing size of the transducers. A second major restraining factor then is the transducer size. Differential pressure across a heart chamber or blood vessel, particularly the veins, would be highly desirable to measure. At the present time there are no differential implantable pressure transducers. The last area of future needs covers the biological compatibility of the transducer material. A nonthrombogenic agent is greatly needed for this purpose. These needs must be fulfilled before the implantable transducer can be routinely used in totally implantable biotelemetry applications. The existing problems represent a drawback for this purpose. Totally implantable telemetry units are needed in such areas of chronic animal research as behavioral studies, toxicology, and physiology. In this manner the functional response of the whole organism can be studied. Implantable pressure transducers must be an initial part of these units.

## FUTURE NEEDS

The use of catheter-tip pressure sensors has been varied; however, in animal implantation studies, the major source of information has been derived from the cardiovascular system.[1–4] Implantable pressure transducers have been used to measure pressures in the right and left sides of the heart – particularly the ventricular chambers – and, at various points, along the aorta. In general, however, most investigators have not used implantable pressure sensors within the low pressure or venous portion of the vascular system. The reasons for this are varied, but it is primarily due to the inability to identify a zero pressure reference in these gauges following implantation. The future needs of the implantable device can be viewed in terms of the current inadequacy of the existing systems. A distinction must also be made between a totally implantable telemetry unit with pressure sensors as opposed to the use of implantable sensors with lead wires that surface through the skin of the animal. The primary comparison in this chapter will be between large and small primates. The future needs for development of implantable pressure devices are predicated on both the current use and limitations of the implantable telemetry unit in particular. The location of implantation is certainly a major criterion in the selection of any measuring device.

For application in a range of animals from the 30-kg chimp[5] to the 3-kg rhesus monkey, a decrease in size and rigidity is needed. The implantable unit shown in Figure 1 has attached pressure transducers which are 6 mm in diameter. This size is difficult to implant in the vascular system of a small monkey weighing 3 kg. In Figure 2, the unit is shown being implanted in a chimp. The left ventricular cell is usually placed through the apex of the heart by a stab wound near the base of the papillary muscles. These larger size transducers also can be placed in the descending thoracic aorta in the chimp without a great deal of trouble, with the exception of the configuration and rigidity of the gauges that have been used. We have found that a much smaller transducer is needed to measure aortic pressure in the monkey. At the present time, a pressure cell of 2 to 3 mm maximum diameter would be optimum for a 3-kg animal. Otherwise, some difficulty is encountered in the long-term placement of these devices. Figure 3 shows an X-ray of a chimp with the implantable unit in the thorax. The placement of the various gauges can be seen. Figure 4 shows this same animal resting in a cage several months after surgery. The batteries and loop antenna in the rear of the cage were used to activate the implantable unit and to deactivate when a recording period was finished.

The problem of transducer size could be ap-

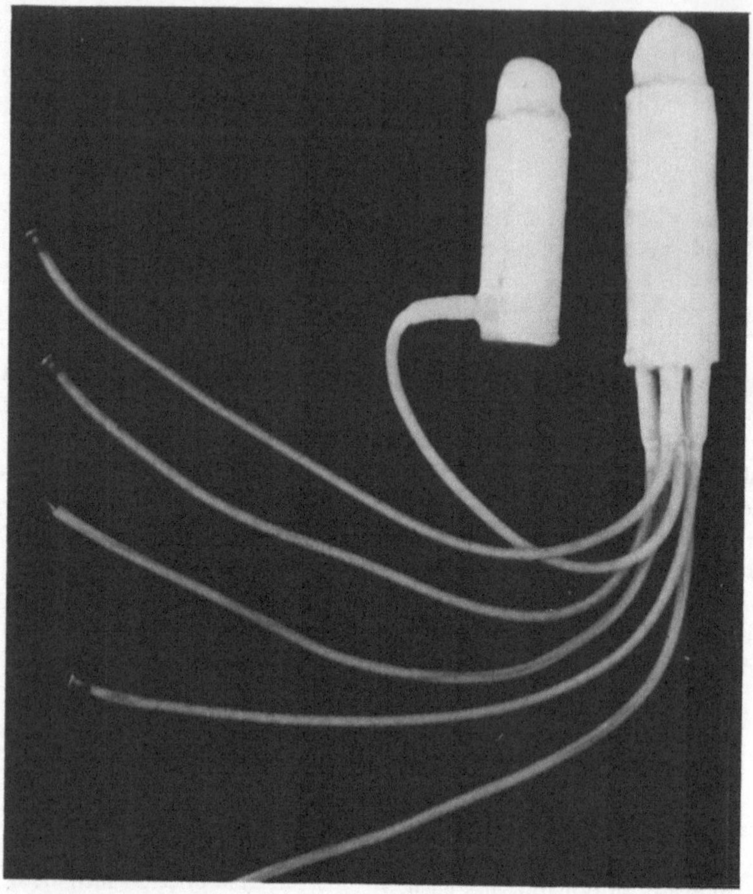

FIGURE 1. Implantable telemetry unit with four pressure cells attached. The unit to the right of the larger unit is the battery power pack. (From Stone, H. L., Sandler, H., and Fryer, T., Implantable telemetry in the chimpanzee, in *Proc Int. Telemetering Conf.*, Vol. 8, International Foundation for Telemetering, Los Angeles, 1972, 470. With permission.)

preciated when the surgical implantation was attempted in a 3-kg rhesus monkey. Figure 5 shows the monkey following completion of the surgery. In this figure, by comparing the size of the surgeon's hand (at left) with the chest incision (in center), it is apparent that there is an order of magnitude difference in size as compared to the chimp. The pressure transducers were implanted in the left ventricle and the thoracic aorta in this animal. Figure 6 shows the implantable unit in place after 6 months. The size of the left ventricular gauge was 6 mm and the aortic gauge was 3.5 mm in diameter. The telemetry unit was placed in the posterior portion of the left thorax. The left ventricular pressure cell can be seen above the unit in the apex of the heart and the aortic pressure cell can be seen just right of the lower edge of the unit.

In comparison to the above implantable units, Figure 7 shows an X-ray of a dog's chest to demonstrate a method having lead wires exit through the external surface of the animal.[6] The wires usually exit the skin on the dorsal surface in the cervical region. Long-term problems associated with this approach to chronic instrumentation are answered by a totally implantable unit.

There are several problems associated with the characteristics of the gauge itself when the transducer is to be implanted for months. The problems are (a) thermal stability and sensitivity; (b) drift; (c) long-term hysteresis and deformation of the gauge; and (d) ability to reliably determine a true zero reference pressure. The last item is really a deterrent to the use of these gauges in certain portions of the vascular system, because of the lack of ability to determine a true pressure. These problems are really design problems, but they

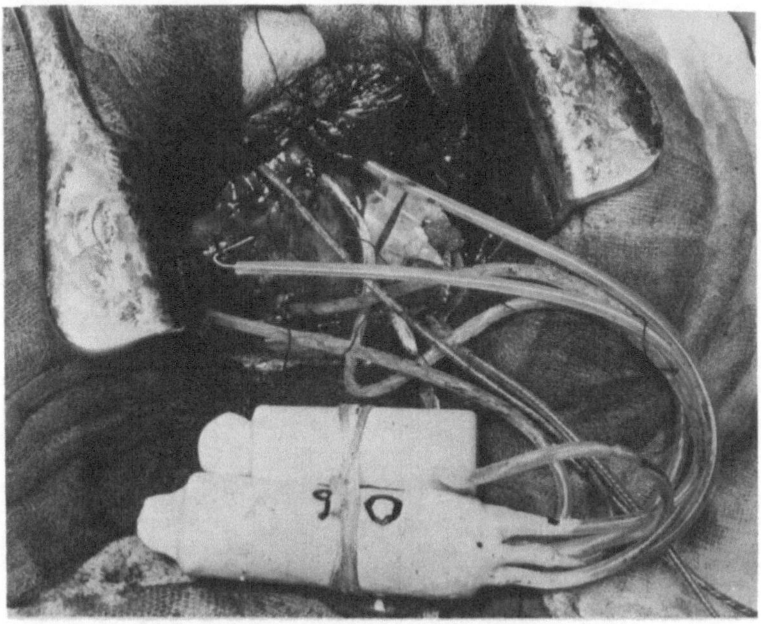

FIGURE 2. Picture taken during implantation procedure of telemetry unit into a chimp. The two units can be seen tied together in the lower center portion of the figure. Note the size of the unit in relation to the size of the blades on a standard rib retractor. (From Stone, H. L., Sandler, H., and Fryer, T., Implantable telemetry in the chimpanzee, in *Proc. Int. Telemetering Conf.*, Vol 8, International Foundation for Telemetering, Los Angeles, 1972, 470. With permission.)

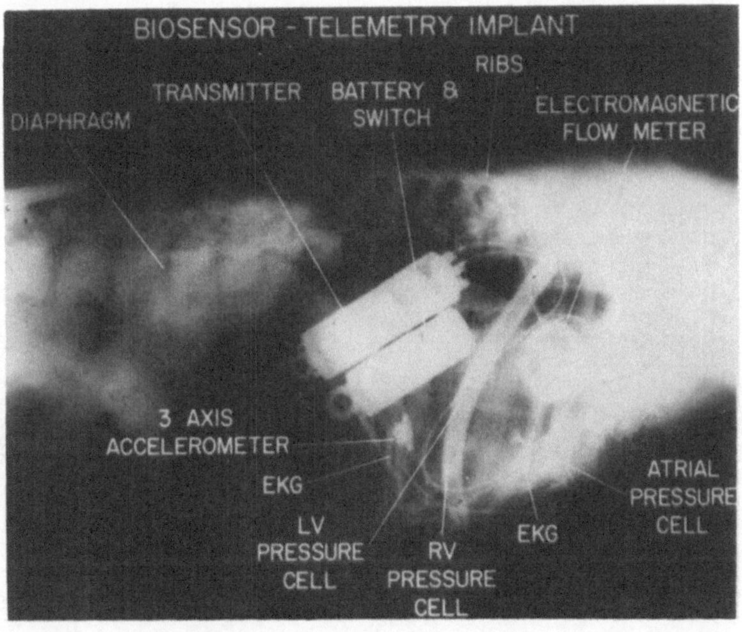


FIGURE 3. X-ray of telemetry implant in the chest of a chimp. The various gauges are listed in the figure.

FIGURE 4. Chimp resting quietly in cage several months following surgery. (From Stone, H. L., Sandler, H., and Fryer, T., Implantable telemetry in the chimpanzee, in *Proc. Int. Telemetering Conf.*, Vol. 8, International Foundation for Telemetering, Los Angeles, 1972, 471. With permission.)

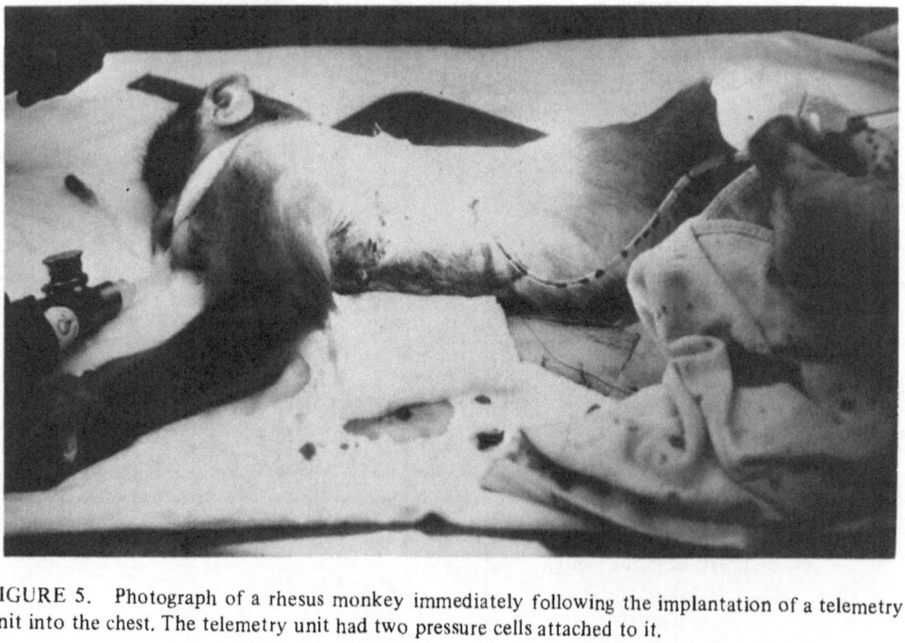

FIGURE 5. Photograph of a rhesus monkey immediately following the implantation of a telemetry unit into the chest. The telemetry unit had two pressure cells attached to it.

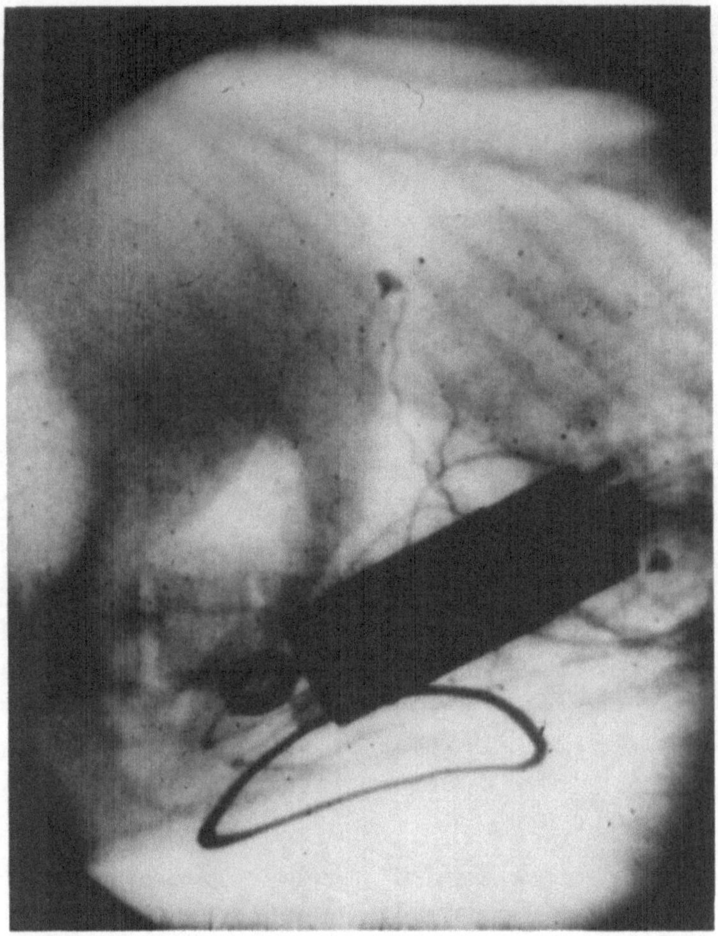

FIGURE 6. X-ray of a rhesus monkey 6 months following implantation of a telemetry unit in the chest.

relate to the long-term use of these devices without some type of cross-reference to a standard pressure measuring system. The method normally used at the present time to determine a true zero reference pressure involves a direct comparison with another pressure signal obtained through a catheter implanted in the animal model or by passing a catheter through an artery to the site of implantation. In certain experiments, this procedure is not desired and may jeopardize the experiment. In the chimp, the pressures in the left ventricle and aorta have been cross-referenced using a catheter-tip sensing device passed retrograde through a femoral artery to the aorta or left ventricle. In the monkey, the same procedure has been accomplished. However, it is very easy to appreciate that for long-term use this procedure has limitations in the ease with which it can be accomplished in the monkey.

The long-term sensitivity of the gauge is an area that can present problems if various techniques for checking the sensitivity are not employed. One method that can be used is in conjunction with the zero reference measures, and that is the infusion of vasoactive agents that will change the pressure. In this way, the sensitivity of the gauge can be determined and compared to the pre-implantation values. A second method, when available, concerns the use of a pressure chamber. Commonly used implantable gauges are sealed at some atmospheric pressure near sea level pressure (760 mmHg). If the barometric pressure is decreased, the measured pressure will shift by an amount equal to the change in barometric pressure.[7] The mmHg change of pressure can be plotted vs. the output from the gauge and compared to the original calibration curve. This particular experiment can be accomplished repeatedly over several months. Small changes in the sensitivity of these particular gauges have been found over this time span.

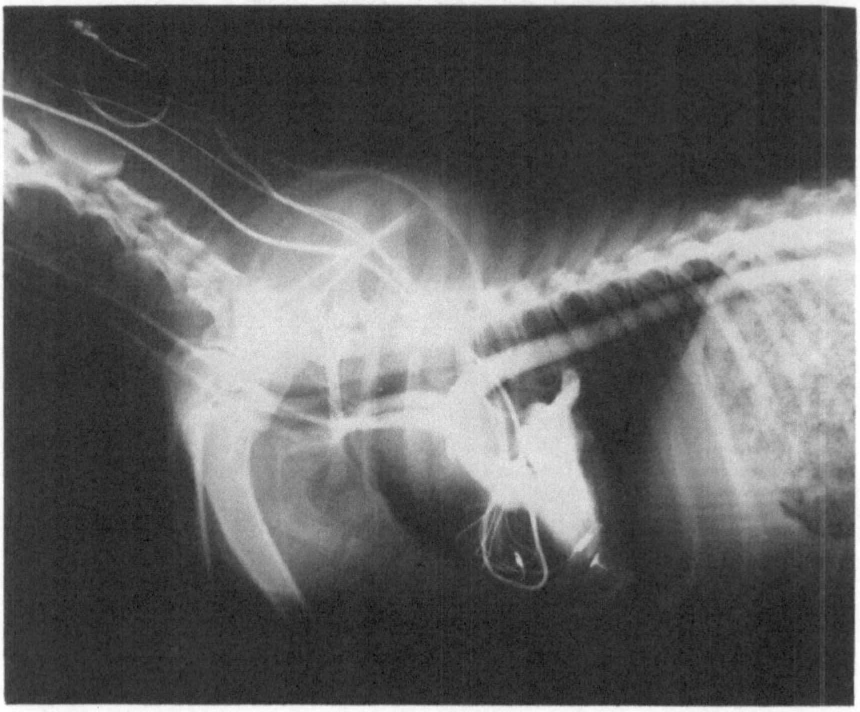

FIGURE 7. X-ray of a dog's chest with the left ventricle and aortic arch visualized with a radiopaque material. The wires of the various transducers can be followed toward the left of the figure along the dog's neck.

Another area of concern is the tissue compatibility of the implantable gauges. An improvement in the thrombogenic properties of the materials used is necessary to prevent long-term changes in the vascular system of animals. Usually the face of the gauges will be coated with some variety of heparin-binding compound that will reduce the formation of clots over the surface of the gauge, particularly in the aorta. In several instances, a streaming clot has been found along the aorta several weeks following the implantation of the transducer. Close examination of peripheral organs revealed that small clots may be continuously dislodged from the surface of the transducer until an endothelial cell proliferation has occurred. Figure 8 shows a gauge implanted through the apex of the left ventricle near the base of one papillary muscle. The hemorrhagic area around the gauge is probably from the stab wound procedure. In this instance, compare the endothelial growth over another device at the left of the figure and that over the pressure transducer. Usually this same type of growth will occur over the pressure transducer in time. The period of this growth process varies with each animal that has been used. In general, after about 2 months, the cells are completely overgrown with this type of tissue. In the aorta, where shear rates are high and the flow along the wall is disrupted by the gauge, this process usually takes a longer period of time.

Materials that would decrease the thrombolitic nature of these devices and promote tissue growth would be highly desirable for long-term use without any complications arising.

The last future need that would be highly desirable is a true differential pressure gauge. There are many experimental conditions in which differential pressure is needed to make interpretation of the data complete. One example of this is an increase in the gravitational force field as shown in Figure 9.[8,9] In particular, notice the left ventricular pressure signal in this record. As the gravitational force is increased, the thoracic pressure also is increased, thus producing an increase in the measured pressure solely from hydrostatic forces. The ideal quantity is the transmural pressure, in this instance, which would be the difference between pleural pressure and chamber pressure. This problem becomes even more acute when one measures the pressure in a low pressure system, such as the atria or veins.

Most users of these gauges assume that the

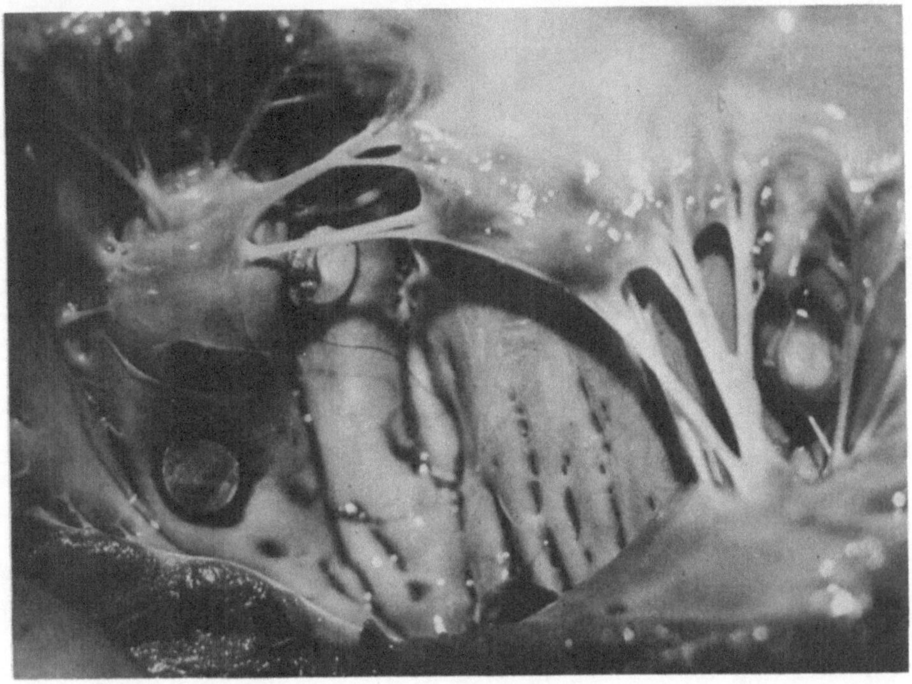

FIGURE 8. The left ventricular cavity was opened to show the implantation of an implantable pressure cell at the apex. Other devices were implanted also and can be seen at the right side of the figure.

extravascular pressure is near zero and will not influence these measurements. But, as seen, there is at least one condition in which this is not true. Other conditions where extravascular pressure will change considerably would include lower body negative pressure, pleural effusion, and external pressure applied over any part of the vasculature.

## APPLICATIONS

The future applications of implantable pressure transducers will be varied, but much more use will be made of these devices in telemetry units which are totally implantable. This type of unit has a wide range of applications in cardiovascular physiology in the chronic animal. Studies in free-ranging animals such as behavioral, toxicological, and space will furnish many answers to questions that, at present, are difficult – if not impossible – to answer. The major reason for using an implantable device is the ability to use it for telemetry. The next stage is a totally implantable unit that has all of the features of an external measuring system but without any wires exiting the skin of the animal model. A very excellent example of this is in toxicology, where long-term studies can be conducted measuring the various pressures without jeopardizing the experiment with external wires or other devices. In space, wires, etc., are a total hazard and the implantable device is the best choice. In both behavioral and free-ranging animal studies, the experiments would be very difficult to perform or interpret without such a device. There is ample evidence that the awake, behaving animal responds differently to many stimuli, as compared to the restrained or anesthetized animal.

Improvement of gauge characteristics will allow the investigation of the neural control of the circulation in the awake, free-ranging animal as well as the determination of possible tissue pressure or even small artery pressure within organs. This capability coupled with flow measurements will furnish invaluable knowledge on the normal reflexes of the animal model and, hopefully, man.

## ACKNOWLEDGMENT

This work was supported in part by NASA Grant NGR 44 088 002.

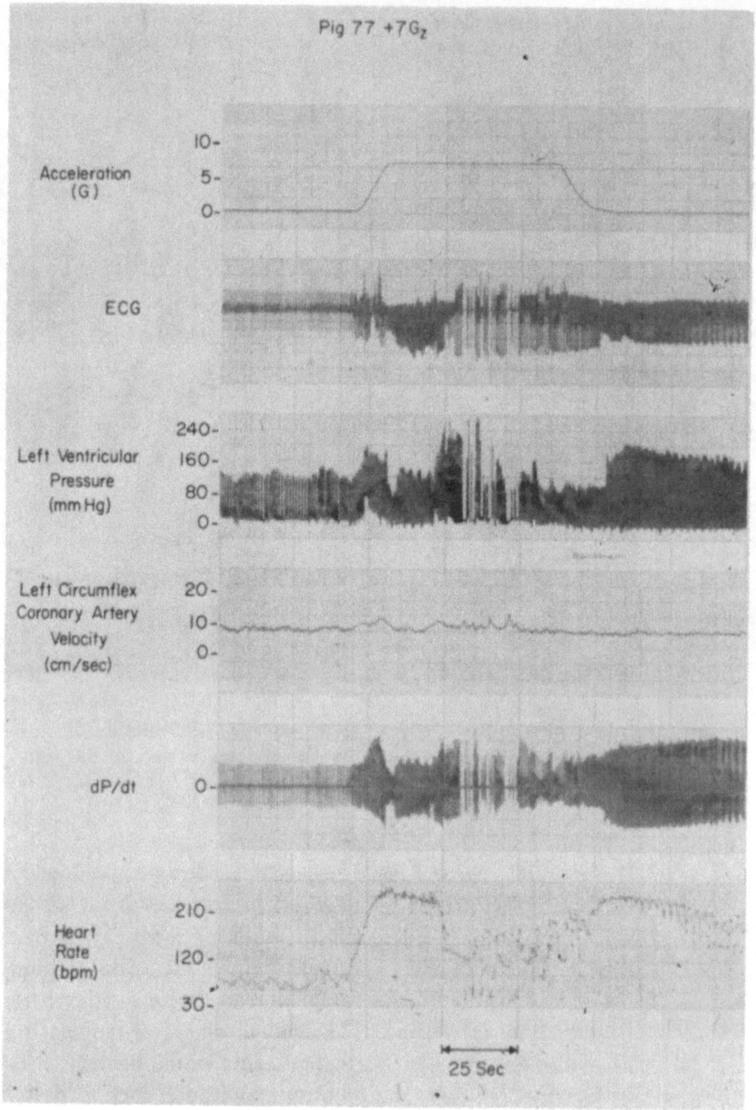


FIGURE 9. Recording of left ventricular pressure, left circumflex coronary artery flow, and heart rate from a miniature swine exposed to 7 $+G_z$. (From Stone, H. L., Sordahl, L. A., Dowell, R. T., Lindsey, J. N., and Erickson, H. H., Coronary flow and myocardial biochemical response to high sustained $+G_z$ acceleration, in *The Pathophysiology of High Sustained $+G_z$ Acceleration, Limitations to Air Combat Manoeuvering and the Use of Centrifuges in Performance Training*, AGARD of NATO Publ., AGARD-CPP-189, Advisory Group for Aerospace Research and Development of the North Atlantic Treaty Organization, London, 1976, A5–7. With permission.)

# REFERENCES


1. **McCutcheon, E. P. and Stone, H. L.**, The significance of chronically implanted instrumentation, in *Chronically Implanted Cardiovascular Instrumentation,* McCutcheon, E. P., Ed., Academic Press, New York, 1973, 3.
2. **Stone, H. L.**, Change in atrial function with cardiac pacing, *Fed. Proc.,* 33(3), 397, 1974.
3. **Adams, J. P., Erickson, H. H., and Stone, H. L.**, Myocardial metabolism during exposure to carbon monoxide in the conscious dog, *J. Appl. Physiol.,* 34, 238, 1973.
4. **Stone, H. L.**, Implantation of solid-state pressure transducers, in *Chronically Implanted Cardiovascular Instrumentation,* McCutcheon, E. P., Ed., Academic Press, New York, 1973, 229.
5. **Stone, H. L., Sandler, H., and Fryer, T. B.**, Cardiovascular function in the chimpanzee studied by chronic instrumentation techniques, in *The Chimpanzee,* S. Karger Medical and Scientific Publishers, Basel, 1972.
6. **Payne, R. M., Stone, H. L., and Engelken, E. J.**, Atrial function during volume loading, *J. Appl. Physiol.*, 31, 326, 1971.
7. **Sandler, H. H., Stone, H. L., Fryer, T. B., and Westbrook, R. M.**, Use of implantable telemetry systems for study of cardiovascular phenomena, *Circ. Res.*, 31, 85, 1972.
8. **Stone, H. L., Erickson, H. H., and Sandler, H.**, Changes in mesenteric, renal and aortic flows with $+G_x$ acceleration, *Aerospace Med.* 45, 498, 1974.
9. **Erickson, H. H., Stone, H. L., Adams, J. P., and Sandler, H. H.**, Coronary blood flow and left ventricular function during environmental stress, in *Proc. Int. Telemetering Conf.*, Vol. 8, International Foundation for Telemetering, Los Angeles, 1972, 206.

# ZERO REFERENCE AND CALIBRATION PROBLEMS OF CIRCULATORY PRESSURES RECORDED BY IMPLANTED AND INDWELLING PRESSURE TRANSDUCERS, AND A LOOK AT FUTURE TECHNOLOGY FOR STUDIES OF DYNAMIC RELATIONSHIPS OF STRUCTURE TO FUNCTION OF THE HEART, LUNGS, AND CIRCULATION*

**Earl H. Wood** and Erik L. Ritman**

The high dynamic response and much reduced sensitivity to motion artifact advantages of miniature indwelling and/or implanted pressure transducers are, for many applications, overshadowed by the technological difficulty of accurate measurements of zero reference levels and differences in pressures recorded simultaneously from different sites in the circulation, by means of these systems. The circulation is a hydrostatic system so that blood pressures at any site in the circulation change 1 cm of water for each centimeter of difference in vertical height within the system. Such a pressure change relative to a given anatomic reference level in, for example, the left or right atria or across a cardiac valve is of considerable hemodynamic and clinical diagnostic importance; hence it should be measurable with reasonable accuracy. Consequently, in addition to knowing what zero pressure is in relation to ambient atmospheric pressure, one must know the exact position of the transducer in relation to a specific fixed anatomic site in the body (e.g., mid-chest level).

The relative ease of obtaining accurate, simultaneous recordings of the zero reference levels and relative sensitivities immediately interchangeable with recording of pressures from multiple sites in the circulatory system is a major advantage of the use of multiple, concentional, externally positioned strain gauge manometer systems for this purpose.

The important deficiencies in the current state-of-the-art of biomedical technology, in relation to significant health care problems such as coronary artery and other types of heart disease, do not involve circulatory pressure measurements. Systemic arterial pressures, particularly, can be measured noninvasively and with sufficient accuracy for clinical purposes by conventional blood pressure cuff methods. Accurate, objective, and quantitative methods for measuring overall and regional myocardial function and perfusion are, however, lacking and badly needed.

Consequently, dynamic computerized spatial tomography of the heart and circulation, rather than pressure transducer developments, should be a very high priority goal in biotechnology research.

*The research upon which this report is based was supported in part by research grants HL04664 and RR-7 from the National Institutes of Health; NCI Contract NCI-CB-53857-33, U.S. Public Health Service; NGR 24-003-001 from the National Aeronautics and Space Administration; and AF 44620-71-C-0069 from the Office of Research, U.S. Air Force.

**Doctor Wood is a Career Investigator of the American Heart Association.

Before discussing some of the problems associated with the use of the currently available miniature pressure transducers, which are progressively becoming more sophisticated and reliable, it is worthwhile to consider the historical development of methods for direct measurement of pressures in the vascular system. This history has evolved over approximately the last 250 years, since the first measurements of arterial pressure in a living mammal were apparently made by Stephen Hales back in 1728 (Figure 1). The era of mechanical recording of blood pressure then lasted for the ensuing two and one half centuries up to about the 1940s. The sensing element of the best mechanical manometer systems at that time consisted of a miniature mirror eccentrically mounted on a flexible but quite stiff diaphragm. An intense light source was focused on the mirror, and the variations in the trace of the resulting light beam caused by pressure-induced movements transmitted by a fluid column to the back side of this diaphragm were recorded on a photokymographic camera.

FIGURE 1. Drawing of Stephen Hales measuring the arterial pressure in a horse. (Commissioned by Statham Instruments, Inc., 1967; *Classics of Cardiology*, Vol. 1, Dover, 1941.)

The Hamilton® manometer[1,2] was one of the first high-frequency mechanical manometers, and it was quite widely used in the 1940s for high-fidelity recordings of arterial blood pressures via a hypodermic needle (Figure 2). It is of historical interest to note that the early recordings of intracardiac pressures in intact humans, which were obtained during the first cardiac catheterization procedures carried out in this country by Cournand and co-workers, were made by using Hamilton manometers.[3]

The current era of electrical recording of blood pressure began in about the mid-1940s with introduction of the unbonded strain gauge manometer, which was perfected by Louis Statham (see Reference 4). This was the first relatively easy to use and practically reliable electrical pressure transducer, and probably for most purposes it is still the method of choice today.[5]

Figure 3 is a picture of an early Statham strain gauge manometer which was adapted for and first used to record arterial blood pressure on the Mayo human centrifuge in the mid-1940s.[6] This was a relatively simple device to use since all that was required was a stable DC voltage source, a high-frequency geophysical-type galvanometer, and a photokymographic recorder. No electronic amplification was required, and this may still be the most reliable method to date for recording of pressures in the vascular system. Since the manometer proper was a relatively low-frequency device, it was necessary, in order to maintain a sufficiently low compliance for adequate recordings of arterial pressure pulses via a hypodermic needle, to use a malleable lead-tube connection between the intra-arterial needle and the manometer.

Figure 4 shows this system as it was used on the Mayo human centrifuge. The subject in this picture is Dr. Edward Lambert, who played a major role in the development of this manometer in our laboratory. A 20-gauge needle rigidly attached to a malleable lead tube was inserted into the radial artery, and the recording site at the wrist was supported at head level so as to measure the variations in arterial perfusion pressure to the brain during the losses in vision (blackout) or consciousness caused by exposure to positive accelerations such as encountered in aerial

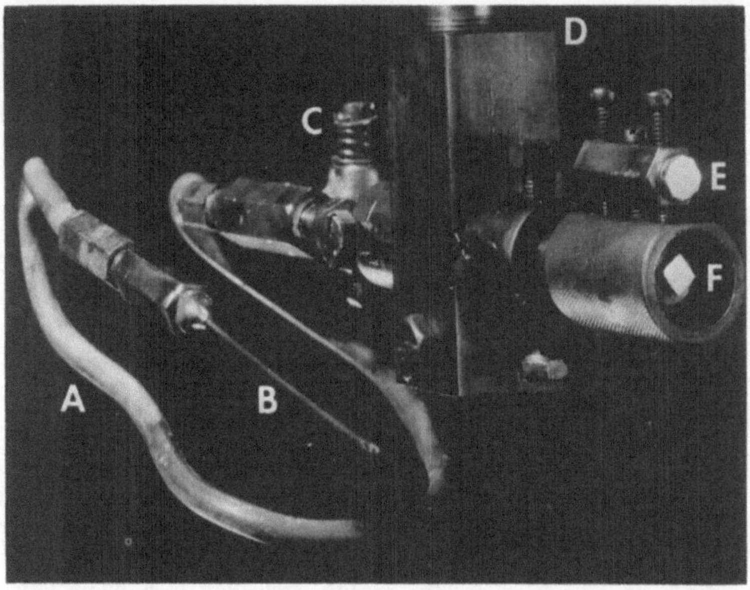


FIGURE 2. Hamilton hypodermic manometer. Malleable lead tubing (A) connecting hypodermic needle (B) via stopcock (C) to the manometer which is fixed in an adjustable position by the rigid mounting assembly (D). The mirror (F) is mounted eccentrically on a flexible copper-beryllium membrane, which is subjected to and flexed by the pressures transmitted to its back surface via the noncompliant, gas-free, liquid-filled hypodermic needle-lead tubing systems. The mirror (E) is rigidly attached to the manometer frame to record any physical movement of the manometer assembly or the imaging light source on the photokymographic recording. Hence, it serves as the base-line trace for measurements of the pressure pulse tracings imaged by mirror F. (Courtesy of Dr. Phillip Dow, Department of Physiology, Medical College of Georgia.)

combat.[6] Figure 5 is a picture of the current version of the Statham® manometer which has evolved on the basis of experiences with the initially successful model.[7]

Figure 6 is a diagram of the Gauer-Gienapp variable reluctance catheter-tip manometer. This, to our knowledge, was the first miniature catheter-tip pressure transducer to be used in human cardiac catheterization procedures.[8] Figure 7 is a picture of this miniature manometer system mounted on the tip of a cardiac catheter, which was taken in 1949 when we were using this device for recording of intracardiac pressures. Initially, when first becoming acquainted with this catheter-tip manometer system, we were very enthusiastic since it appeared to avoid many of the frequency response problems and catheter motion artifacts routinely present in conventional cardiac catheter recordings. However, the zero reference and calibration problems associated with its use overshadowed its advantages so that, except for special purposes, we have avoided whenever possible the use of catheter-tip and implantable pressure transducers ever since.

This fact highlights a very important advantage of the use of conventional externally positioned manometer systems. The relative ease of obtaining accurate, simultaneous recordings of the zero reference levels and the relative sensitivities of multiple conventional strain gauge manometer systems immediately interchangeable with recordings of pressures from multiple sites in the circulatory system is illustrated by Figure 8. These simultaneous traces of radial artery, aortic, and right ventricular pressures were recorded in immediate temporal sequence with simultaneous *in situ* recordings of each manometer's zero reference level, and the deflections were produced by each system by identical pressures of 50 and 100 mm of mercury referenced to the same anatomic level. In this instance, as was frequently the case, the midvertical height of the chest at the

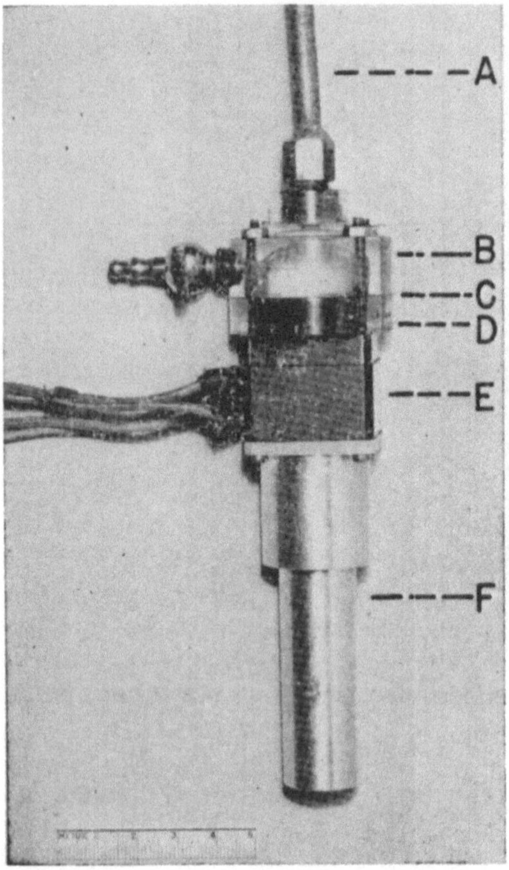


FIGURE 3. Early Statham strain-gauge manometer adapted for measurement of arterial blood pressure. A. Lead tube to which hypodermic needle is attached (see Figure 4). B. Lucite® chamber filled with anticoagulative solution via stopcock and fluid reservoir. C. Plastic membrane. D. Lucite chamber filled with oil. E. Case containing strain-gauge elements connected by wire leads to battery and galvanometer. F. Holder. (From Lambert, E. H. and Wood, E. H., *Proc. Soc. Exp. Biol. Med.*, 64, 186, 1947. With permission.)

level of the juncture of the third interspace with the sternum was used as the anatomic zero reference level.[7]

Interpretation of the physiologic or pathologic significance of pressure recordings from multiple sites in the circulation requires: (1) accurate knowledge, to within 1 cm of water, of the possible differences in pressure between two sites (e.g., upstream and downstream to a cardiac valve) at each instant in time throughout individual cardiac cycles, and (2) the zero reference level of each manometer system in relation to a given anatomic site in the circulation.

Due to the major physiological importance of these two types of pressure measurements, the critical problem of obtaining these data with high accuracy when using indwelling pressure transducers should be considered. Actually, the technological difficulty of accurate measurements of the zero reference levels and absolute differences in pressures recorded simultaneously from different sites in the circulation, by means of indwelling miniature pressure transducers, is the major problem in the use of these systems. From the viewpoint of circulatory pathology or physiology, it is not enough to know what zero pressure is in relation to ambient atmospheric pressure, but rather it should be possible to measure blood pressures, particularly on the venous side of the circulation, with an accuracy of at least 1 cm of water in relation to a zero reference level at a specified anatomic site in the body. This is because the circulation is a hydrostatic system so that blood pressures at any site in the circulation increase or decrease by 1 cm of water for each centimeter of difference in vertical height within the system. A pressure change of 1 cm of water in relation to a given anatomic reference level in, for example, the left or right atria or across a cardiac valve may be of considerable hemodynamic and clinical diagnostic importance; hence, it should be measurable with reasonable accuracy.[7]

The relatively simple and accurate means of establishing the zero reference levels and relative sensitivities of multiple external manometer systems is illustrated in Figure 9. The two manometer systems can, at will, be connected via plastic tubing and stopcocks to the hydraulic system shown in this figure, which contains sterile heparinized Ringer's solution. The meniscus of this solution in the inverted glass reservoir (wash bottle) of the system, which is suspended on a rack-and-pinion gear, can be set at any desired anatomic level. Meaningful measurements of central circulatory pressures in a subject that remains stationary in the supine or prone positions require that the anatomic zero reference level be determined by a lateral thoracic roentgenogram or the externally measured ventral-dorsal diameter of the chest.[7,9] The air pressure above the meniscus in the reservoir bottle is adjustable to any desired level by means of a conventional hand bulb and monitoring mercury manometer (Figure 9). The procedure of obtaining zero reference levels is more difficult during changes in body position or exposures to changes in the direction or magnitude

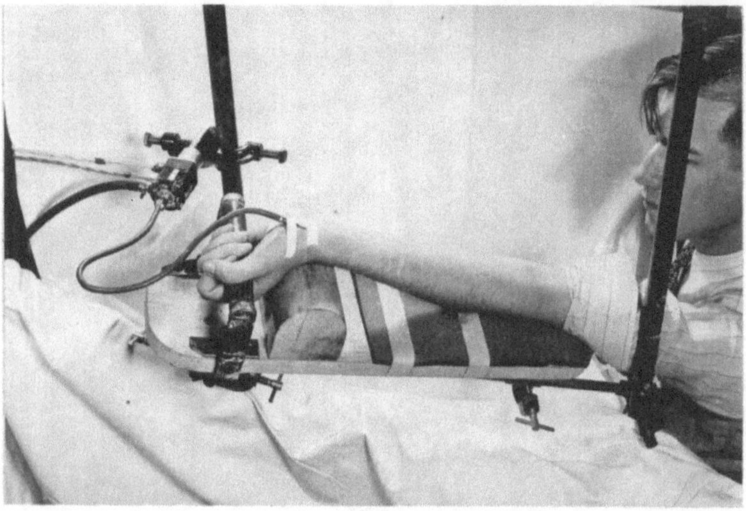

FIGURE 4. Hypodermic strain-gauge manometer assembly and arm board support assembly for rigid positioning in the cockpit of the Mayo human centrifuge for recording of arterial pressure at brain level during exposure to positive (+ $G_z$) acceleration (see Figure 3 for detailed view of manometer). The arm board support fixes the wrist in a dorsiflexed position to facilitate percutaneous puncture of the radial artery.[7,23] (From Lambert, E. H. and Wood, E. H., *Proc. Soc. Exp. Biol. Med.*, 64, 186, 1947. With permission.)

of the gravitational-inertial force environment, such as that which occurs on animal or human centrifuges and in aerospace flight.[10,11]

An assembly such as that shown in Figure 9 allows simultaneous recording of the zero reference levels of multiple manometer systems at any time and as frequently as desired.[7,9-11] This is accomplished by opening the air chamber above the wash-bottle meniscus to ambient atmospheric (zero) pressure, and by opening all of the manometers to this hydrostatic system via the appropriate stopcocks and liquid-filled tubing system. Zero reference levels established in this manner are independent of the position of the manometers. Therefore, the individual manometers can be fixed in any position in relation to the subject that is most convenient, even on the floor and/or the ceiling, if so desired, providing the relative positions of the subject and the manometers are not changed during the recording and calibration procedure.

In addition to their respective zero reference levels, the relative sensitivities of multiple manometer systems can be established very accurately, and at any desired instant in time, by recordings of the deflections produced simultaneously by all manometers when they are connected via the hydraulic-tubing stopcocks system to the wash-bottle reservoir, when its air chamber has been adjusted to the desired known increments in pressure. These simultaneous recordings of stepwise changes from the multiple physiologic pressures to identical constant incremental calibration pressures in all gauges provide very accurate determinations of their relative sensitivities, as well as an indication of the frequency response and possible hysteresis characteristics of each manometer system (Figure 8).[7] Visualization of the problem and the physiologic importance of establishing the anatomic zero reference level of recordings of intracardiac pressures obtained from indwelling or implanted miniature pressure transducers can be facilitated by reference to Figure 10, which includes a drawing of a cross-section of a dog thorax at mid-heart level. The internal diameter of, for example, the left atrium of a large dog may be 5 cm. Consequently, the pressure recorded from a miniature manometer within this chamber may vary 5 cm of water, depending on the position of the manometer within this chamber when there is no change in atrial pressure. This type of artifactual change in pressure, which will be recorded if a change in the position (i.e., the vertical height)

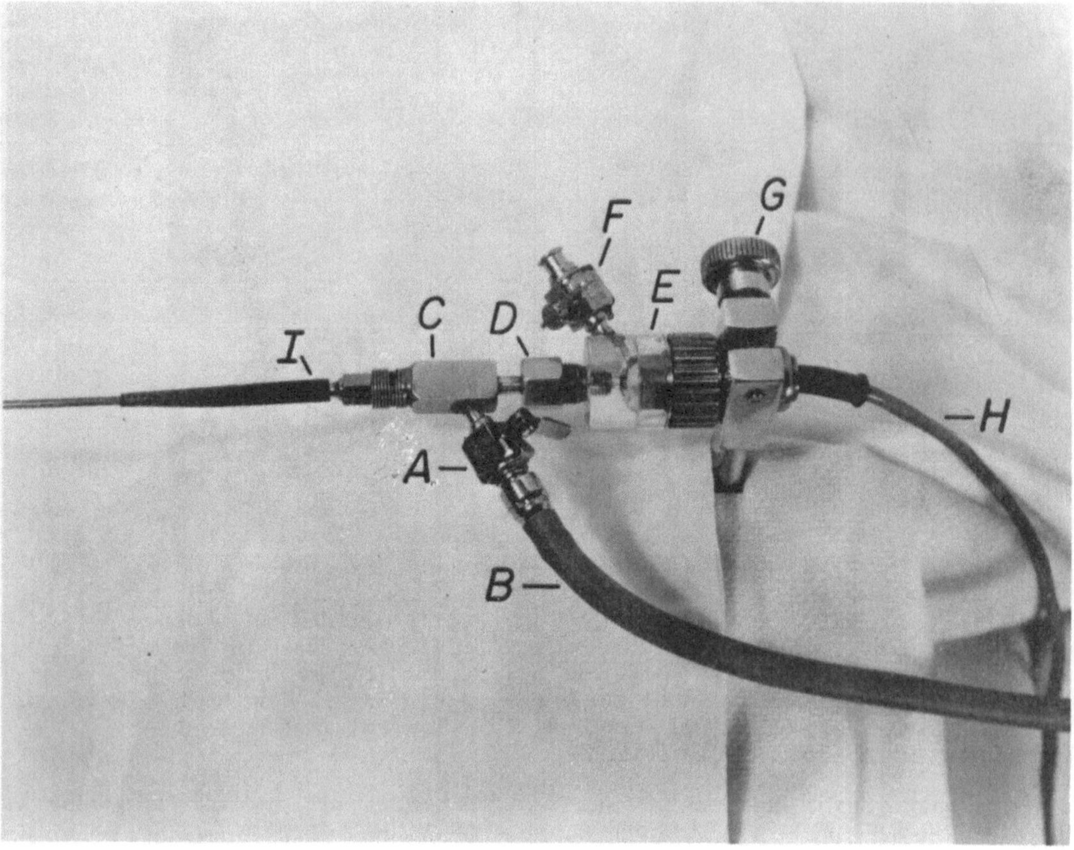


FIGURE 5. Modern strain-gauge manometer (Statham type P23D) with flushing by-pass assembly (C) and cardiac catheter (I) attached. Stopcock (A) is for the intermittent flushing of catheter system with sterile Ringer's solution contained in tubing (B) leading to pressurized wash-bottle reservoir (see Figure 9 for details). C is connected to strain gauge (E) by a special Luer-Lok® connection reinforced by locking nut (D). Accessory stopcock F is used to flush sterile saline solution through manometer flushing head, when system is assembled, to obtain an air-free system. G is the mounting bracket for strain gauge and H is the 4-wire electrical lead connecting strain gauge to DC power supply and oscillographic galvanometer. The flushing by-pass assembly (C) is an essential component for use of current Statham transducers for accurate recordings of blood pressure. This is because prolonged variations in the zero base line of the system caused by the transient upset of the steady-state temperature gradients in the Wheatstone bridge sensing element diaphragm assembly occur if the system is flushed with a sterile solution via stopcock F. (From Sutterer, W. F. and Wood, E. H., Strain-gauge manometers: Application to recording of intravascular and intracardiac pressures, in *Medical Physics,* Vol. 3, Glasser, O., Ed., Year Book Medical, Chicago, 1960, 641. With permission.)

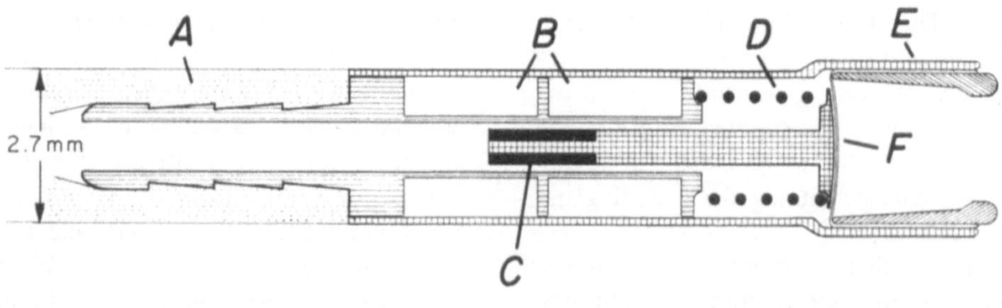


FIGURE 6. Diagram of Gauer-Gienapp cardiac catheter-tip manometer. A. Number 8F cardiac catheter. B. Housing for primary and secondary coils of variable reluctance sensing transformer. C. Soft iron slug directly coupled to backside of diaphragm (F). D. Metal coil spring. E. Protective shield for pressure-sensitive diaphragm (F). (Reproduced with permission of Dr. Otto Gauer, Department of Physiology, Free University of Berlin).

of a catheter-tip manometer in the great vessels or cardiac chambers occurs, does not result when a conventional external manometer-cardiac catheter assembly is used, since the pressures recorded by this system are independent of the position of the catheter tip in a hydrostatic system. In the case of miniature transducers implanted in a fixed position in the atrial wall, the recorded pressure could change 5 cm of water when the dog rolled over when the atrial pressure had, in fact, remained constant.

Correction for this type of artifactual change in pressure as well as interpretation of the physiologic significance of the absolute level of pressures recorded by a miniature indwelling or implanted pressure transducer require accurate knowledge of the position of the manometer in relation to a specified anatomical reference point. In case of measurements in intact animals or human beings, this requires visualization and measurement of the position of the miniature transducer from biplane roentgenograms as illustrated in Figure 10. Since there is also a vertical gradient in pericardial and pleural pressures[9-12] and the position of the heart in the thorax changes with variations in body position,[11,12] these factors also have to be accounted for in relation to estimates of transmural and effective cardiac-filling pressures based on recordings from indwelling transducers. This is true particularly during changes in body position in free-ranging animals or during exposures to changes in direction or magnitude of the gravitational-inertial force environment.[11]

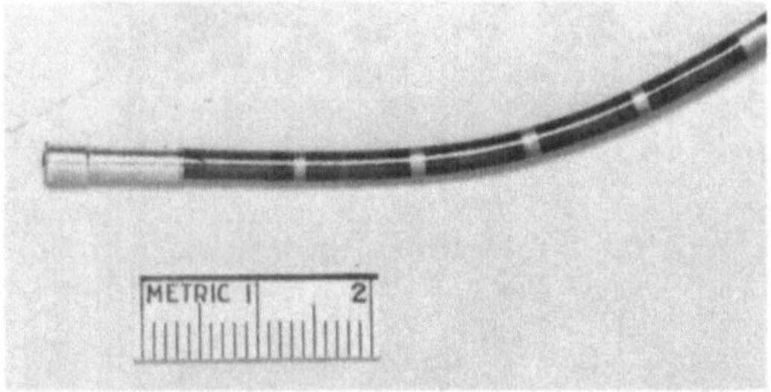


FIGURE 7. Gauer-Gienapp intracardiac manometer attached to an 8F Cournand® catheter. (From Ellis, E. J., Gauer, O. H., and Wood, E. H., *Circulation*, 3, 390, 1951. With permission.)

One inherent advantage of indwelling miniature manometers is their high-frequency response. However, since (as illustrated in Figures 11 and 12) optimally damped systems with a uniform

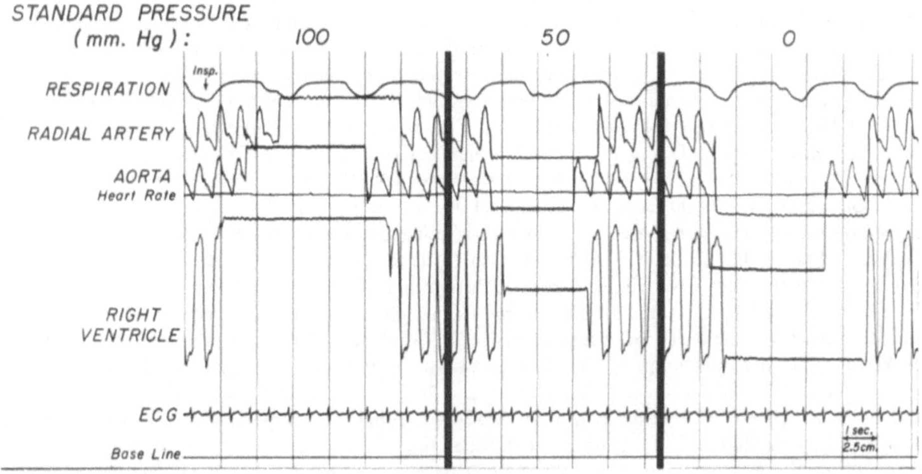


FIGURE 8. Example of recordings of *in situ* calibration of multiple manometer systems using calibration system shown in Figure 9. Recordings of multiple pressures are obtained immediately before and after each simultaneous stepwise exposure of manometers to decrements of standardizing pressures used. By this means, errors are minimized that may result from inadvertent changes in sensitivity, shifts in base line, or hysteresis-like effects in individual manometers. Furthermore, each manometer system is automatically referenced to the same zero pressure level. This method of calibration is strongly recommended when highly accurate comparisons of simultaneous pressures at multiple sites in the circulatory system are required. (From Sutterer, W. F. and Wood, E. H., Strain-gauge manometers: Application to recording of intravascular and intracardiac pressures, in *Medical Physics*, Vol. 3, Glasser, O., Ed., Year Book Medical, Chicago, 1960, 641. With permission.)

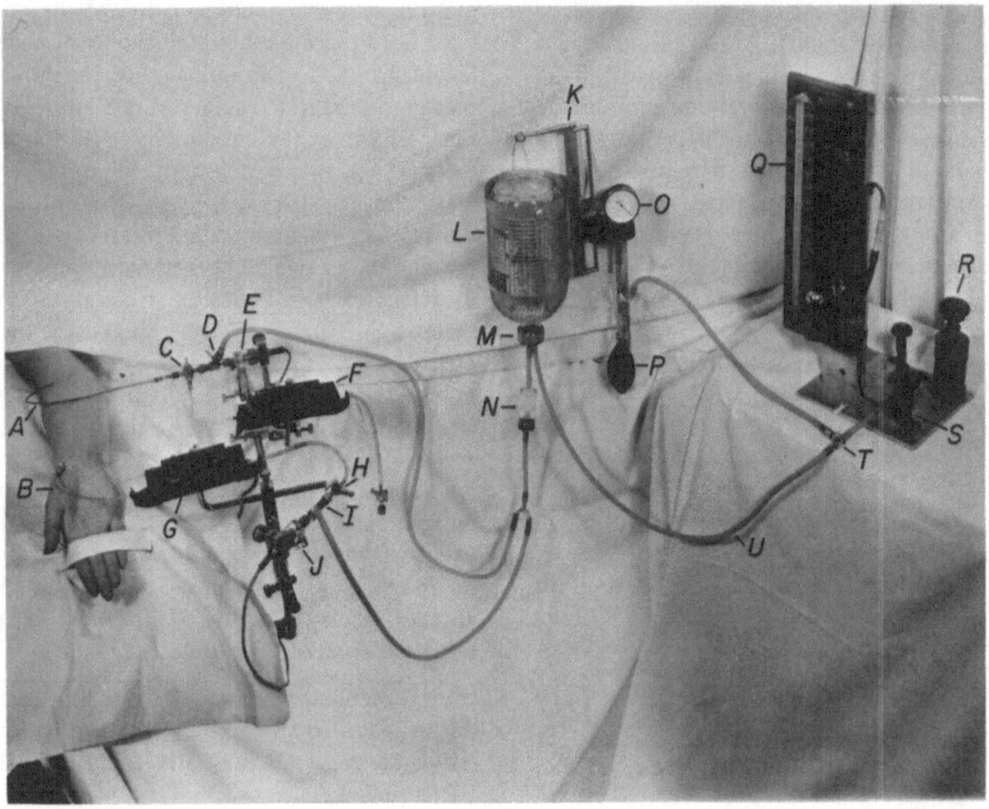


FIGURE 9. Assembly for calibration of zero reference levels and sensitivities of multiple manometers *in situ* (used in this laboratory for diagnostic right-heart catheterization[25]). The two strain-gauge manometer systems are incorporated in an assembly attached to a patient for continuous interchangeable recording of intracardiac plus systemic arterial pressures, $O_2$ saturation, or indicator concentration of blood withdrawn from these sites. Catheter (A) to right side of heart can be connected, via two-way stopcock (C), to strain-gauge manometer (E) via its flushing by-pass assembly (D), or to cuvette oximeter (F) for continuous recording of $O_2$ saturation or indicator concentration in blood being withdrawn from the catheter. (see Figure 5 for construction of D and E.) Indwelling needle (B) in radial artery is connected to strain-gauge manometer (J) via cuvette oximeter (G). A two-way stopcock (H) allows interchangeable withdrawal of arterial blood samples via its side arm or recording of arterial pressure via the flushing by-pass assembly (I). Pressurized wash bottle (L) with protective mesh-wire covering is suspended from rack-and-pinion gear mount (K), which facilitates adjustment of meniscus of wash bottle to zero reference level (see text). L is a one-liter intravenous bottle (Baxter®) containing Ringer's solution and 20 mg sodium heparin. It is converted to a pressurized wash bottle by the threaded adapter cap (M) which incorporates a connection between the most dependent fluid level in the bottle and the strain gauges, via a tubing system, float valve (N), and the one-way stopcocks on respective side arms or the flushing by-pass assemblies of the gauges. The float valve (N) minimizes the possibility of accidental infusion of air into the patient should the wash bottle be inadvertently emptied of fluid. Air chamber at top end of wash bottle is connected to pneumatic calibrating system via tubing (U) and 2-way stopcock (T). During routine operation and flushing of strain-gauge systems, pressure of about 300 mmHg is maintained in this air chamber by means of hand bulb (P) and Tycos® manometer assembly (O). For calibration of manometers, meniscus of wash bottle is first adjusted to mid-thoracic level, and pressure in the air chamber is accurately adjusted to desired standardizing pressure by screw-controlled pneumatic piston (R), mercury manometer (Q), and needle valve vent (S). Manometers are then exposed to this standard pressure by suddenly opening stopcocks D and I and simultaneously closing stopcocks C and H. For recordings of such a calibration, see Figure 8. (From Sutterer, W. F. and Wood, E. H., Strain-gauge manometers: Application to recording of intravascular and intracardiac pressures, in *Medical Physics*, Vol. 3, Glasser, O., Ed., Year Book Medical, Chicago, 1960, 641. With permission.)

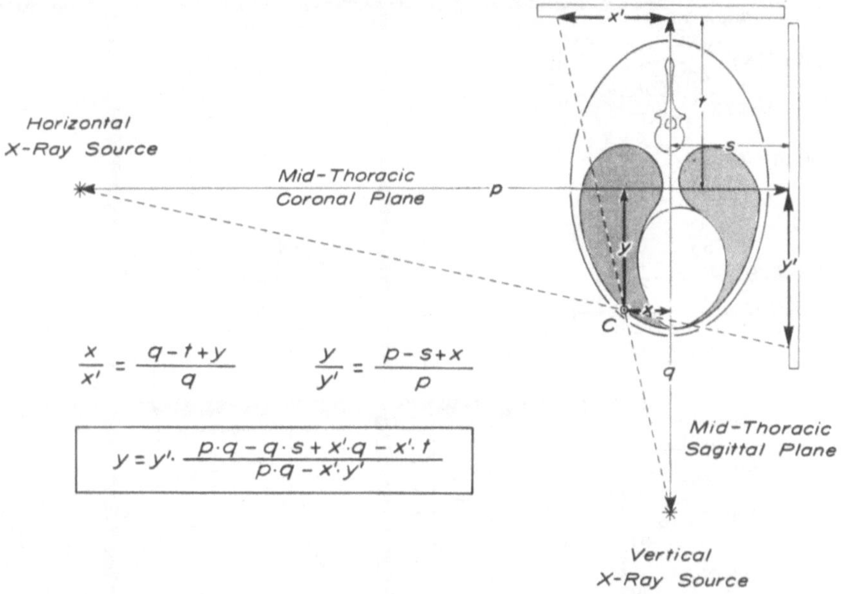


FIGURE 10. Illustration of method used to determine the position of a catheter tip (C) relative to the mid-chest planes by means of biplane roentgenograms. The vertical distance of the projection of the catheter tip on the plate from the mid-thoracic coronal plane is indicated by y', x' is the horizontal distance of its projection from the mid-sagittal plane, taking the sign into consideration. The vertical distance of the catheter from the mid-thoracic coronal plane for supine or prone positions is given by the formula in the insert. (From Rutishauser, W. J., Banchero, N., Tsakaris, A. G. Edmunowicz, A. C., and Wood, E. H., *J. Appl. Physiol.*, 21, 1500, 1969. With permission.)

amplitude response and time lag from zero to 10 to 20 Hz are adequate for recording of the practically significant components of most intravascular pressures,[7] this advantage of miniature manometers is frequently overemphasized. Given a uniform dynamic response from zero to 10 to 40 Hz, the ease and absolute accuracy of measurement of pressures in relation to a known and fixed anatomical site and the ability to measure differences in pressures between various sites in the circulatory system with an accuracy of better than 1 cm of water are of much greater physiologic utility than is a high-frequency response.[7]

The important deficiencies in the current state-of-the-art of biomedical technology in relation to significant health care problems (such as cardiovascular, especially coronary artery, disease and various types of heart failure) do not involve circulatory pressure measurements. Systemic arterial pressures, particularly, can be measured noninvasively and with sufficient accuracy for clinical purposes by conventional blood pressure cuff methods.

What is really needed are more accurate, objective, and quantitative methods for measuring overall and regional myocardial function and perfusion. The chemical to mechanical energy conversion capabilities of the myocardial muscle cell are at the basis of all cardiac functions. The cyclic changes in the shape, dimensions, and tension of the myocardial walls, which are required to maintain the circulation, are generated by changes in length and tension of the small cylindrical cardiac muscle cells which are the major components of the myocardium. Consequently, to study the regional and overall function of these cells, it is necessary to have an accurate method to measure not only the pressures in and around the heart, but also dynamic changes in its shape and dimensions. The required pressure measurements are within the realm of current technology. Unfortunately, the required dynamic regional and overall shape and dimension measurements are not and can never be obtained from conventional roentgenographic projection imagery or ultrasound techniques.[13]

The recent development of computerized cross-sectional reconstruction techniques

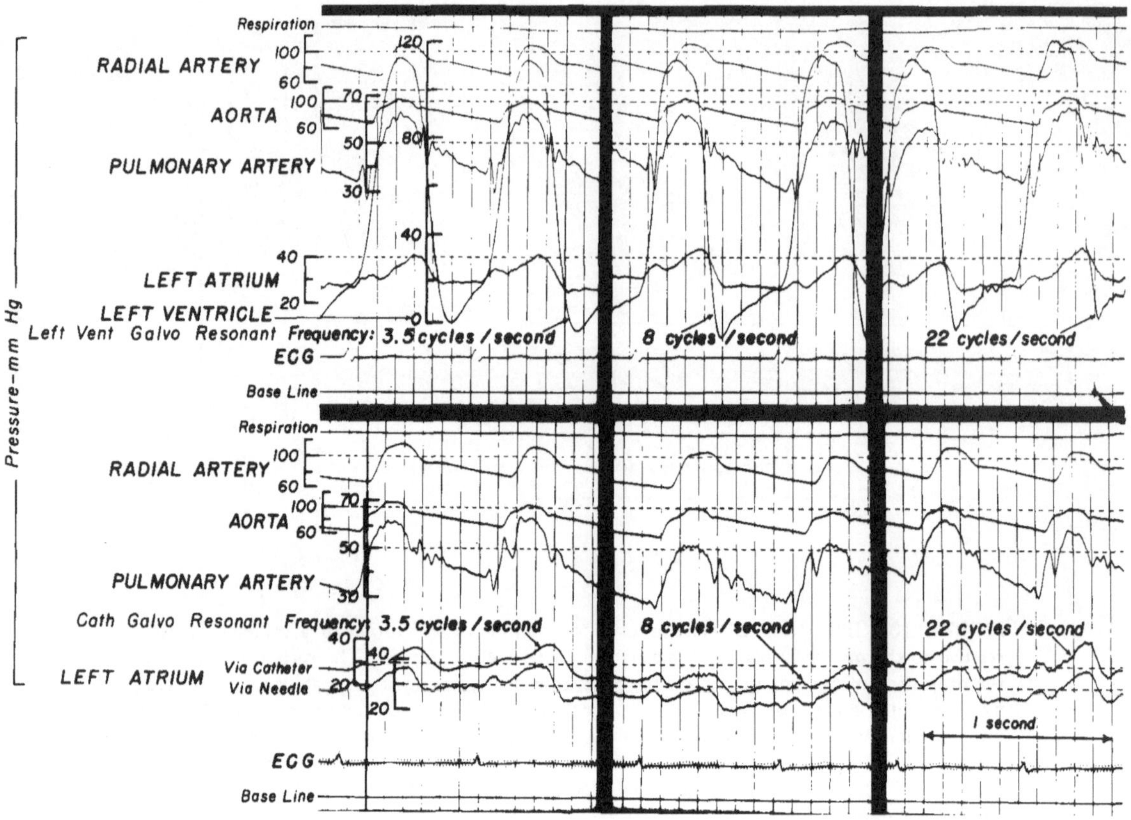


FIGURE 11. Effect of variation in dynamic response of manometer recording system on contours of left ventricular (above) and left atrial pressure pulses (below) in a 41-year-old woman with mitral insufficiency and chronic constrictive pericarditis. Recordings were obtained in rapid succession with different manometer systems, designated according to the natural frequency of the recording galvanometer used (3.5, 8, and 22 Hz). For overall dynamic response curves of these systems, see Figure 12. Vertical lines delineate 0.1-sec intervals; camera speed was 150 mm/sec. Minimal and maximal ventricular pressures are not significantly different, although upper range of uniform response to sine-wave variations in pressure for the three systems was 3.5 to 22 Hz (Figure 12). Damping and time delay of pressure pulses recorded by the 3.5-Hz system are most obvious in the recording of left atrial pressure pulses (below, left), although minimal and maximal left atrial pressures are indicated with reasonable accuracy. No appreciable difference was noted between the left atrial pressure pulses recorded simultaneously by the 22-Hz catheter and 40-Hz left atrial needle systems (below, right). Dynamic response of these two systems is similar to that used routinely for simultaneous recording of left ventricular and left atrial pressures via a single 18T-gauge needle assembly introduced percutaneously into the left atrium.[24] (From Marshall, H. W., Woodward, E., Jr., and Wood, E. H., *Am. J. Cardiol.*, 2, 24, 1958. With permission.)

(computerized axial tomography) provides the potential capability of obtaining accurate measurements of the shape and dimensions of the full anatomic extent of the cardiac chambers at up to 60-per-second instants in time throughout individual cardiac cycles.[14] However, because of the lack of temporal resolution imposed by their dependence on mechanical scanning and their limited anatomical axial scanning range and resolution, current commercially available brain or whole-body scanners are not capable of accurate dynamic reconstructions of the epi- and endocardial surfaces of the heart or the coronary circulation of intact animals or human beings.[15]

It has been demonstrated under conditions of computer-controlled physiologic stationarity of the cardiac and respiratory cycles (i.e., identical amplitude, phase, and position reproducibility of heart beats in constant phasic relationships with the respiratory cycle throughout successive breaths, during the mechanical and hence prolonged scanning period) that 60-per-second three-dimensional reconstructions of the parietal pleural surfaces and the epi- and endocardial surfaces covering the full anatomic extents of the heart and thorax are possible in dogs, using a single X-ray source image intensifier television system.[16] Such a system requires electronic x-y (planar) transla-

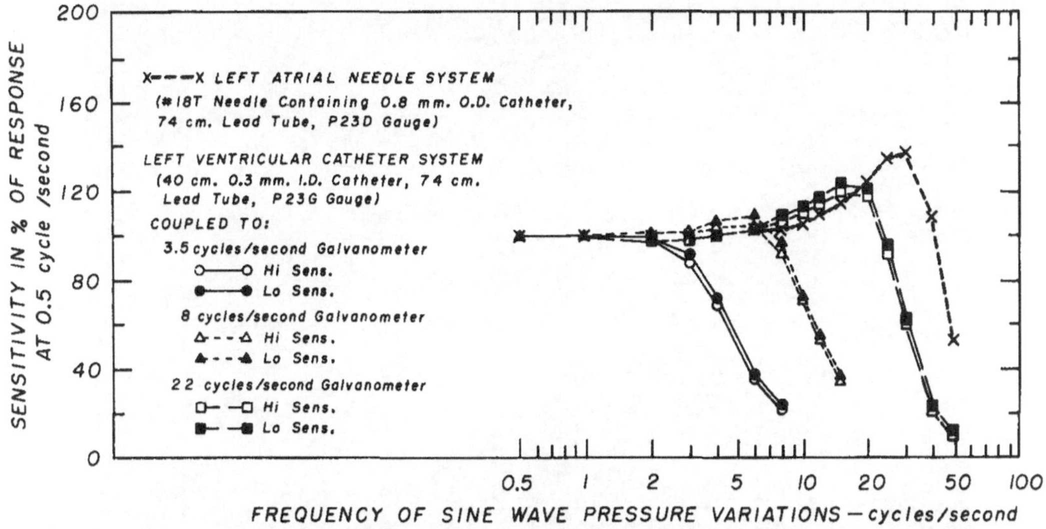


FIGURE 12. Dynamic response of manometer systems used to study effect of variation in dynamic response of manometer systems on left ventricular and left atrial pressure pulses recorded by these different systems. Identical manometer systems were used to record left ventricular and left atrial pressure pulses shown in Figure 11. (From Marshall, H. W., Woodward, E., Jr., and Wood, E. H., *Am. J. Cardiol.*, 2, 24, 1958. With permission.)

tional scanning with constant phasic relationships between each 60-per-second scan plus the cardiac and respiratory cycles. In addition, the mechanical axial (circumferential) scanning must be synchronized with these events and timed so as to cover a complete respiratory cycle at each incremental angle of view over a range of 180° or more.[17]

Figure 13 is a montage of computer-generated television images of 16 parallel, *temporally synchronous,* cross-sections of the thorax of an intact dog, produced by Richard Robb and colleagues of the Mayo Biophysical Sciences Unit. Multiple reconstructed cross-sections of this type, which cover the full dorsal-ventral and right-to-left dimensions of the thorax concomitantly with a simultaneous axial anatomic range large enough to encompass the full cephalad-to-caudad dimensions of the thorax plus a high degree of axial resolution, require a very high planar scanning speed of the full thorax (i.e., 250 lines per 60-per-second video field) from which multiple temporally synchronous cross-sections can be computed.[18]

This type of scanning and subsequent three-dimensional reconstruction provides three-dimensional matrix arrays of the X-ray digital attenuation values (voxels) encompassing the full thorax. Consequently, these data can be used to reconstruct selected or complete sets of synchronous parallel coronal or sagittal sections encompassing the full thorax. These or other sections oriented at any desired angle in relation to the x, y, and z axes of the thorax can be computed at any or all desired 60-per-second instants in time throughout the cardiac or respiratory cycles.[19]

Unfortunately, because of the impracticality of producing physiologic stationarity of the rate, amplitude, and phasic relationships of the cardiac and respiratory cycles in unanesthetized animals or human beings, this type of potentially valuable, basic investigative and clinical diagnostic data can never be obtained by currently operational or envisaged mechanical brain or whole-body scanning systems.[18]

A new generation high temporal resolution, simultaneous high axial anatomic range and axial resolution, and all-electronic planar and axial tomographic scanning system is required to provide these types of heretofore unobtainable data. Such a system has been designed and can be fabricated using commercially available components.[18,20] This system (Figure 14) will be capable of dynamic high axial anatomic range and resolution multiple synchronous cross-sectional reconstructions of the full anatomic extent of continuously moving organ systems (particularly the heart and lungs), including the three-dimensional vascular anatomy and perfusion dynamics in all regions of the body.[18]

The quantitative dynamic spatial imaging capabilities provided by fully electronic whole-

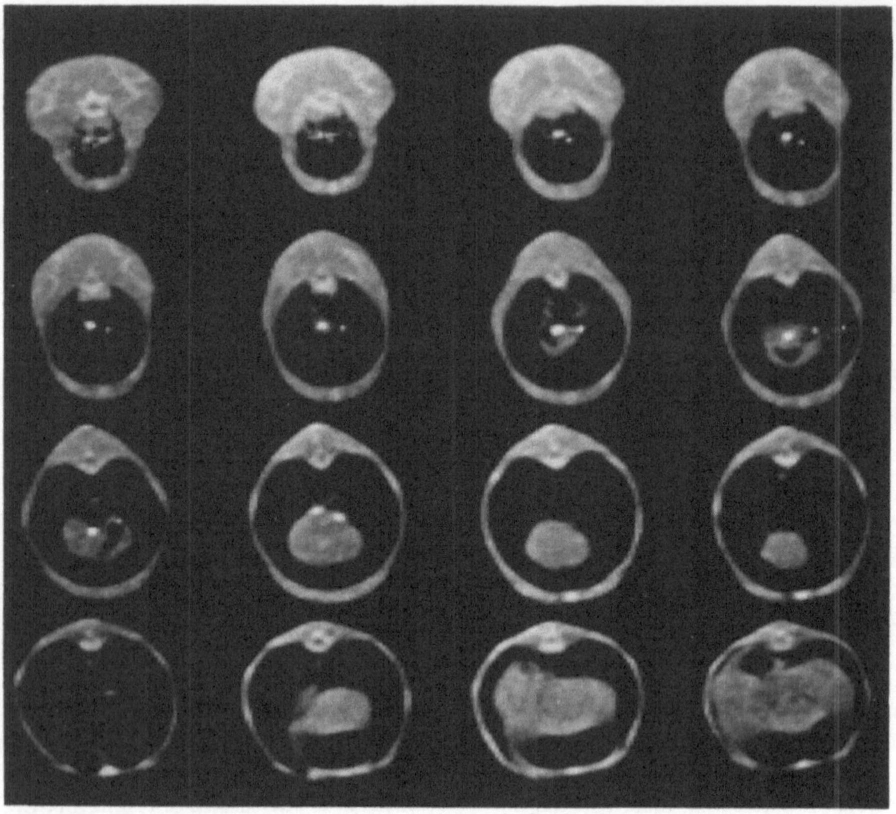

FIGURE 13. Sixteen *synchronous* cross-sections reconstructed over the anatomic extent of the thorax of an intact dog beginning at the apex of the lungs (upper left) and extending left-to-right and top-to-bottom to the diaphragmatic margin at the base of the lungs (lower right). Cross-sections are 1.5 cm apart, selected from a set of 64 cross-sections 3 mm apart (every fifth cross-section used in this figure). Each cross-section was reconstructed from density profiles obtained by digitizing and averaging four successive video lines at each level (3 mm thick) for each of 35 views of the thorax in 6° increments over a total range of 204°. Note that the spinal vertebra, ribs, and three (2-mm diameter) catheters in the heart stand out as bright regions. The pleural, epicardial, and diaphragmatic surfaces are well delineated as well as part of the bowel, which contains some trapped air and protrudes cranioventrally in relationship to the cephalad extent of the dome of the diaphragm and underlying liver (bottom row of cross-sections). Such a complete set of multiple *synchronous* parallel cross-sections encompassing the full anatomic extent of the thorax cannot be obtained by the current first and second generation EMI®, ACTA®, and DELTA® whole-body scanners because of their lack of temporal resolution, and inadequate anatomic axial range and axial resolution.[18-21] The resulting complete three-dimensional matrix of voxels of X-ray attenuation values encompassing the entire thorax provides the capability of rapid computation of selected or complete sets of synchronous high axial resolution parallel coronal or sagittal sections encompassing the full anatomic extent of the heart or lungs or oriented at any desired angle in relation to the x, y, and z axes of the body. (From Robb, R. A., Ritman, E. L., Greenleaf, J. F., Sturm, R. E., Liu, H. K., Chevalier, P. A., and Wood, E. H., *J. Assoc. Comput. Mach.*, 10, 240, 1976. With permission.)

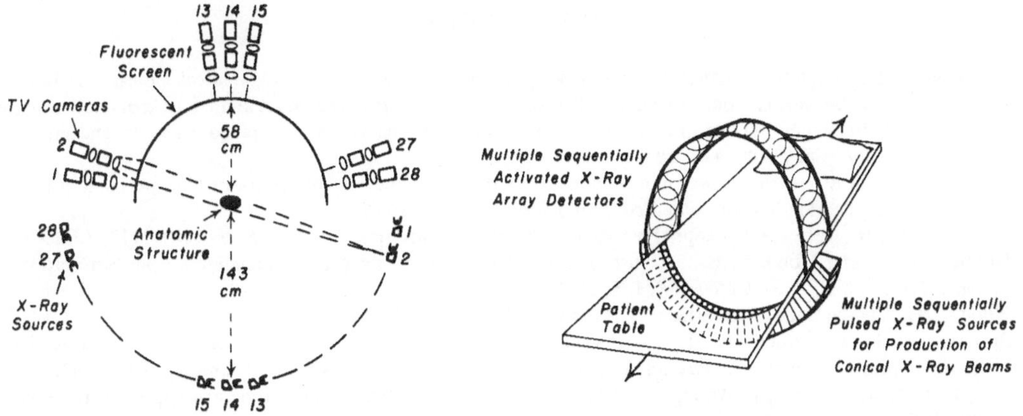


FIGURE 14. Diagram of a proposed prototype of a general purpose, fully electronic scanning, high temporal resolution and synchronous high axial anatomic range and resolution dynamic spatial reconstruction system (DSR).[18] This system utilizes 28 X-ray sources paired with the same number of videoroentgenographic detectors arranged in respective semicircles in the same plane as the desired cross- sectional reconstruction (i.e., for studies of the heart and lungs, the cephalocaudad axis of the chest would be perpendicular to this plane and be interposed between the opposing semicircular arrays of X-ray sources and detectors). The X-ray beam from each X-ray source is projected through the anatomic structure onto a segment of a semicircular fluorescent screen which is coupled optically to an image-intensifier/video chain somewhat similar to those used in standard fluoroscopic procedures.[20] The individual components of this system are commercially available and in current use in radiology or for other technologic applications. (From Wood, E. H., *Chest*, 69, 394, 1976. With permission.)

body tomographic scanning systems of this type, coupled with multiple high-fidelity pressure measurements using currently available (and continuously improving) external indwelling and implantable miniature pressure transducers, will allow quantitative studies of the static and dynamic relationships of three-dimensional anatomic structure and circulatory function in all organ systems of the body, which have been impossible heretofore. Such studies will increase the understanding and clinical diagnostic acumen of many types of clinically important disease processes. They will also add a very powerful, new and needed tool applicable to a myriad of important and currently unresolved biomedical investigative, clinical diagnostic, and health care problems.[21]

Thus, support of dynamic quantitative spatial reconstruction imaging techniques and pressure transducer developments should be included as a very high priority goal in biotechnology research. These developments will provide new horizons for increased understanding and consequent improved care of the heart, lungs, and circulation in all regions of the body, in health and disease.[15,21]

## ACKNOWLEDGMENT

The authors are indebted to their professional and technical colleagues in the Mayo Biophysical Sciences Unit, upon whose work and thinking this review is largely based.[22]

# REFERENCES


1. **Hamilton, W. F.,** Circulation, in *Medical Physics*, Vol. 1, Glasser, O., Ed., Year Book Medical, Chicago, 1944.
2. **Hamilton, W. F., Brewer, G., and Brotman, I.,** Pressure pulse contours in the intact animal. I. Analytical description of a new high frequency hypodermic manometer with illustrative curves of simultaneous arterial and intracardiac pressures, *Am. J. Physiol.*, 107, 427, 1934.
3. **Cournand, A.,** Cardiac catheterization, development of the technique, its contributions to experimental medicine, and its initial application in man, *Acta Med. Scand. Suppl.*, 579, 3, 1975.
4. **Comroe, J. H., Jr.,** Retrospectroscope: Hydrogen, balloons and pressures, *Am. Rev. Respir. Dis.*, 113, 73, 1976.
5. **Lambert, E. H. and Wood, E. H.,** The use of a resistance wire strain gauge manometer to measure intraarterial pressure, *Proc. Soc. Exp. Biol. Med.*, 64, 186, 1947.
6. **Wood, E. H.,** Charlie Code: Reminiscences, *Mayo Clin. Proc.*, 50, 497, 1975.
7. **Sutterer, W. F. and Wood, E. H.,** Strain-gauge manometers: Application to recording of intravascular and intracardiac pressures, in *Medical Physics*, Vol. 3, Glasser, O., Ed., Year Book Medical, Chicago, 1960, 641.
8. **Ellis, E. J., Gauer, O. H., and Wood, E. H.,** An intracardiac manometer: Its evaluation and application, *Circulation*, 3, 390, 1951.
9. **Coulam, C. M. and Wood, E. H.,** Regional differences in pleural and esophageal pressures in head-up and head-down positions, *J. Appl. Physiol.*, 31, 277, 1971.
10. **Sass, D. J., Nolan, A. C., and Wood, E. H.,** Digital computer analysis of circulatory and respiratory pressures in water-immersed dogs breathing liquid in force environments of 1 and 7 Gy, *Aerosp. Med.*, 45, 1, 1974.
11. **Avasthey, P. and Wood, E. H.,** Intrathoracic and venous pressure relationships during responses to changes in body position, *J. Appl. Physiol.*, 37, 166, 1974.
12. **Rutishauser, W. J., Banchero, N., Tsakiris, A. G., and Wood, E. H.,** Effect of gravitational and inertial forces on pleural and esophageal pressures, *J. Appl. Physiol.*, 22, 1041, 1967.
13. **Wood, E. H., Ritman, E. L., Sturm, R. E., Johnson, S. A., Spivak, P., Gilbert, B. K., and Smith, H. C.,** The problem of determination of the roentgen density, dimensions and shape of homogeneous objects from biplane roentgenographic data with particular reference to angiocardiography, in *Proc. San Diego Biomedical Symp.*, Vol. 11, San Diego Biomedical Symposium, San Diego, 1972, 1.
14. **Gordon, R., Herman, G. T., and Johnson, S. A.,** Image reconstruction from projections, *Sci. Am.*, 233, 56, 1975.
15. **Wood, E. H.,** Cardiac and pulmonary dynamics by quantitative imaging, *Circ. Res.*, 38, 131, 1976.
16. **Robb, R. A., Greenleaf, J. F., Ritman, E. L., Johnson, S. A., Sjostrand, J. D., Herman, G. T., and Wood, E. H.,** Three-dimensional visualization of the intact thorax and contents: A technique for cross-sectional reconstruction from multiplanar X-ray views, *Comput. Biomed. Res.*, 7, 395, 1974.
17. **Robb, R. A., Wood, E. H., Ritman, E. L., Johnson, S. A., Sturm, R. E., Greenleaf, J. F., Gilbert, B. K., and Chevalier, P. A.,** Three-dimensional reconstruction and display of the working canine heart and lungs by multiplanar X-ray scanning videodensitometry, in *Computers in Cardiology*, IEEE Computer Society, Long Beach, California, 1974, 151.
18. **Ritman, E. L., Sturm, R. E., and Wood, E. H.,** Needs and performance requirements of spatial reconstruction methods for diagnostic and investigative studies of cardiopulmonary and circulatory dynamics, in *Reconstructive Tomography in Diagnostic Radiology and Nuclear Medicine*, Ter-Pogossian, M. M., Phelps, M. E., Brownell, G. L., Cot, J. R., Jr., Davis, D. O., and Evens, R. G., Eds., University Park Press, Baltimore, in press, 1976.
19. **Robb, R. A., Ritman, E. L., Greenleaf, J. F., Sturm, R. E., Liu, H. K., Chevalier, P. A., and Wood, E. H.,** Quantitative imaging of dynamic structure and function of the heart, lungs and circulation by computerized reconstruction and subtraction techniques. Proceedings of the SIGGRAPH Conference, July 14–15, 1976, Philadelphia, Pennsylvania, *J. Assoc. Comput. Mach.*, in press, 1976.
20. **Sturm, R. E., Ritman, E. L., Johnson, S. A., Wondrow, M. A., Erdman, D. I., and Wood, E. H.,** Prototype of a Single X-ray Imaging Chain Designed for High Temporal Resolution Computerized Tomography by Means of an Electronic Scanning Dynamic Spatial Reconstruction System, *Proc. San Diego Biomedical Symp.*, February 4–6, 1976, San Diego, California, Academic Press, New York, in press, 1976.
21. **Wood, E. H.,** New horizons for study of the cardiopulmonary and circulatory systems, *Chest*, 69, 394, 1976.
22. **Johnson, S. A., Robb, R. A., Greenleaf, J. F., Ritman, E. L., Gilbert, B. K., Storma, M. T., Sjostrand, J. D., Donald, D. E., Herman, G. T., Sturm, R. E., and Wood, E. H.,** Dynamic three-dimensional reconstruction of beating heart and lungs from multiplanar roentgen-television images, *Mayo Clin. Proc.*, 49, 958, 1974.
23. **Wood, E. H., Lambert, E. H., Baldes, E. J., and Code, C. F.,** Effects of acceleration in relation to aviation, *Fed. Proc.*, 5, 327, 1946.
24. **Wood, E. H., Sutterer, W., Swan, H. J. C., and Helmholz, H. F., Jr.,** Technic and special instrumentation problems associated with catheterization of left side of the heart, *Proc. Staff Meet. Mayo Clin.*, 31, 108, 1956.
25. **Wood, E. H.,** Special technics of value in cardiac catheterization laboratory, *Proc. Staff Meet. Mayo Clin.*, 28, 58, 1953.

# IMPLANTABLE PRESSURE GAUGES FOR USE IN CHRONIC ANIMALS*

**Hank Baig, Thomas A. Patrick, and Stephen F. Vatner**

Solid state pressure transducers are routinely used for chronic measurement of arterial and left ventricular pressures in a variety of mammalian species, including primates. This type of device offers a number of advantages over the conventional pressure-measuring system which consists of a fluid-filled catheter and strain gauge manometer. Of principle importance are superior frequency response and suitability for telemetry applications. The major disadvantage of the solid state gauge is the lack of calibration in terms of absolute levels of pressure. Furthermore, the electrical characteristics of the gauges tend to drift with time, thereby requiring application of cross-calibration procedures when the implantable gauges are used chronically.

## INTRODUCTION: CHARACTERISTICS OF IMPLANTABLE PRESSURE TRANSDUCERS

The miniature implantable pressure transducer device most commonly used in our laboratory consists of a small, hermetically sealed (7-mm diameter, 1-mm thickness), cylindrical chamber with a titanium diaphragm, on the back of which are bonded four silicon strain gauge elements arranged in a Wheatstone bridge configuration.** The transducer senses pressure changes as a deformation of the diaphragm, which unbalances the bridge and generates an offset signal proportional to the applied pressure. This signal is subsequently amplified and recorded as an analog reproduction of the pressure wave. These gauges, which were originally developed by Van Citters and Franklin,[1] have been used in our laboratory for several years in mongrel dogs, adult sheep, lambs, and primates, primarily to obtain measurements of left and right ventricular and aortic pressures, with the animals conscious and unrestrained.[2-4]

This kind of pressure transducer offers a number of advantages over commonly used fluid-filled catheter systems for use in chronic implants. The most important of these are frequency response and suitability for telemetry.

A pressure gauge must have a high-frequency response in order to accurately reproduce the rapidly changing phases of a pressure signal, such as the leading edge of systole in left ventricular pressure. Furthermore, the time derivative of left ventricular pressure is often used as a measure of cardiac contractility, and to obtain this parameter by analog differentiation, the higher frequency components of the pressure signal must be preserved (Figure 1). The upper frequency limit of the implanted solid state gauge is 1.2 kHz, as compared with approximately 65 Hz for a commonly employed strain gauge manometer*** which is used in conjunction with fluid-filled catheters. The actual frequency response of the catheter-manometer system in chronic animal preparations is usually much lower due to clotting or kinking of the implanted catheter.

There are several features of the implanted gauge which make it preferable for use in experiments involving radiotelemetry of pressure data. These include both mechanical and electrical aspects of the solid state pressure transducer. Fluid-filled pressure transducers are larger in size and are quite fragile. Movement of the animal may cause a disturbance of the fluid in the dome of the transducer, or in the catheter itself, which can induce acceleration artifacts in the pressure signal. The fluid-filled transducer has a voltage output which is three to four times lower than that of a solid state gauge. This imposes stricter requirements on the design of the associated electronic circuitry necessary for signal conditioning and telemetry. Furthermore, the lower bridge im-

*Supported in part by U.S. Public Health Service Grants HL 15416 and 17459, NASA Grant NSG 2136, and American Heart Association Grant 76907.

**Konigsberg P22, Konigsberg Instruments, Pasadena, California.

***Statham® P23 Db, Statham Instruments, Oxnard, California.

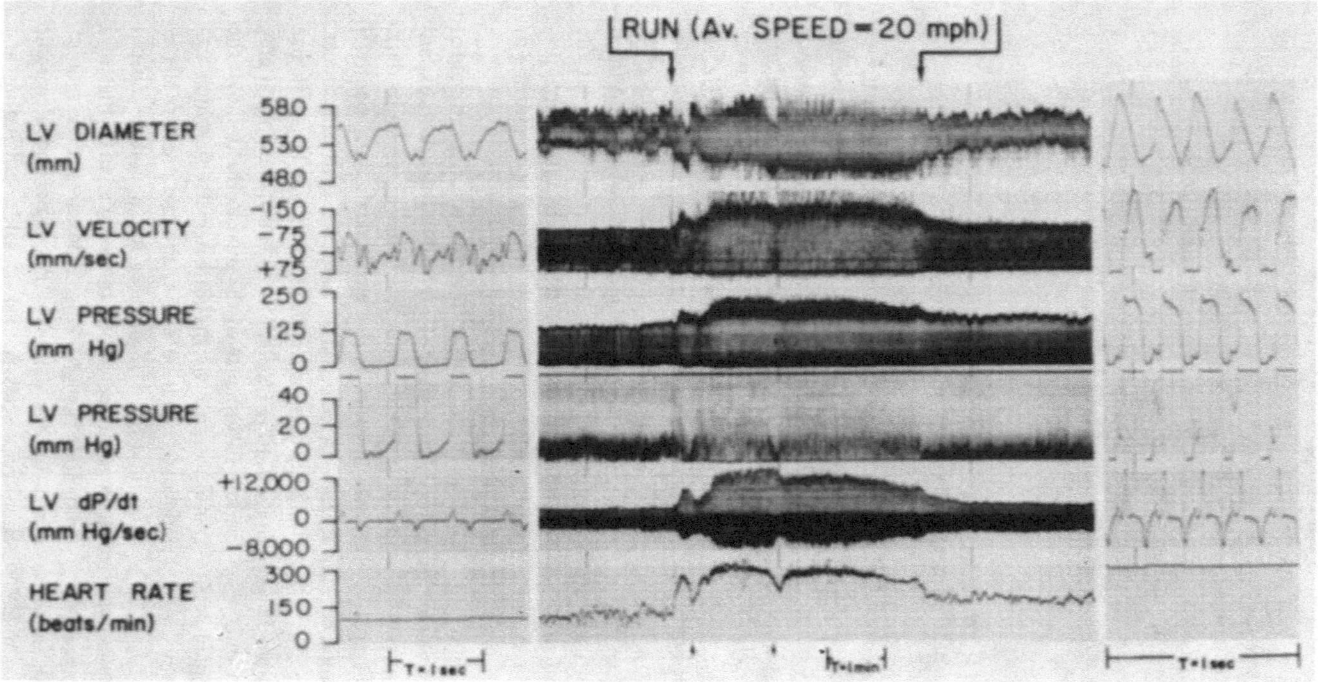


FIGURE 1. A typical response in a dog of phasic left ventricular diameter (epicardial), velocity, pressure, diastolic pressure, dP/dt, and heart rate measured and telemetered during severe exercise. The pressure signal was obtained from a solid-state implanted gauge. (From Vatner, S. F., Franklin, D. L., Higgins, C. B., Patrick, T., and Braunwald, E., *J. Clin. Invest.*, 51, 3052, 1972. With permission.)

pedance of the fluid-filled gauge causes the resultant power requirement to be two to three times greater than that of a high impedance solid state gauge. This is of importance when pressure data must be collected for long periods of time. In addition, catheters tend to clot, and when the animal is inaccessible, hemorrhage or infection may occur.

For these reasons, i.e., power requirements, size, stability, and maintenance, the implantable pressure gauges are well-suited to investigations in which the mobility of the animal is of primary importance, such as studies of severe exercise in dogs (Figure 2), behavioral studies, or experiments with larger free-ranging primates such as baboons, chimpanzees[4] (Figure 3), and giraffes.[5]

The major disadvantages of implantable pressure gauges for use in chronic implants are drift in zero offset and sensitivity. These problems become critical because it is impossible at the time of the experiment to calibrate the implanted gauge in terms of absolute pressure. In addition, since the unit is sealed at the factory, the output is sensitive to changes in barometric pressure, which cause large offset at altitude but also result in significant variations at sea level with normal fluctuations in barometric pressure. Therefore, it becomes critical to cross-calibrate these gauges with another system, either catheter and strain gauge or calibrated catheter-tip manometer.

## IN VITRO CALIBRATION AND MAINTENANCE OF IMPLANTABLE PRESSURE TRANSDUCERS

Each transducer undergoes an initial in vitro calibration prior to surgical implant, and additional calibrations are performed immediately upon removal from each animal in order to determine the extent of zero offset and sensitivity drift. These factors can be described by a linear equation:

$$\text{Output Voltage (mV)} = A + B \times \text{Pressure (mmHg)}$$

where A is the zero pressure offset voltage (mV), and B is the transfer sensitivity (mV/mmHg), i.e., the change in output voltage per mmHg change in pressure.

To calibrate the gauge in vitro, the probe is sealed in a small hermetic chamber which is directly connected to a mercury manometer. The proper excitation voltage, typically 5 V in our laboratory, is applied and the output of the gauge

FIGURE 2. A healthy dog running behind the mobile recording unit in an unrestrained exercise experiment. The dog is wearing saddle-bags which contain a pressure/flow telemetry system.

FIGURE 3. A conscious unrestrained chimpanzee with a backpack containing a battery-operated pressure/flow telemetry system.

is measured with a digital voltmeter at selected pressure levels: 0, 75, 150, 200, 225, and 300 mmHg. Sensitivity is determined from a least-squares regression analysis of these data points.

Next the probe is removed from the chamber and immersed to a depth of 2.54 cm in a controlled temperature water bath. The temperature is varied from 33 to 42°C while the output is measured at constant pressure to obtain a measure of temperature stability, defined as the change in output voltage per degree of change in temperature, with pressure held constant. This parameter causes few problems since the transducer environment is relatively constant with temperature, and fixed resistors external to the gauge element serve to compensate the unit for the expected variations in temperature.

When the zero pressure voltage offset has changed more than ±10 mV from its initial value, the pressure gauge is rebalanced by changing one of the resistors in the external compensation module, returning the offset to its nominal value of 0 mV. The gauge is returned to the factory for repair if: (a) there is any sign of unstable or intermittent operation, or (b) there are severe changes in any of the primary electrical characteristics, i.e., sensitivity, zero pressure offset, temperature stability, or bridge resistance.

The connector end of the transducer lead suffers from fatigue failures as well as catastrophic separation usually caused by the instrumented animal. When this occurs, color-coded copper extension wires are soldered to the transducer leads to retain use of the gauge. It must be pointed out, however, that the resultant change in lead length effects the total bridge resistance and may cause an imbalance leading to an appreciable zero offset, which must be rebalanced as previously described.

## DRIFT IN SENSITIVITY AND ZERO OFFSET

In order to quantitatively assess drift in zero offset and sensitivity, calibration data from 35 pressure gauges used in our laboratory over the past 4 years were examined. Over a period of approximately 80 weeks, zero offset fell by 5.14 ± 0.70 mV, and sensitivity fell from 95.25 ± 4.97 to 78.92 ± 3.96 mV/mmHg (Figure 4). Changes in zero offset and sensitivity are not necessarily linear with time. This point is best illustrated by examining the repeated calibration data for two individual gauges (Figures 5 and 6). The broken lines in Figures 5 and 6 indicate the times the gauges were returned to the manufacturer for recalibration for the reasons previously explained above.

In order to obviate the problems associated with drift, it is necessary to periodically calibrate the gauges in vivo. By doing so, the drift in implanted gauges can be followed and eliminated as a source of error (Figure 7).

## IN VIVO USE AND CALIBRATION

During an experiment the gauge is excited by a constant DC voltage source (5.0 V), and the offset voltage (typically 0 to 20 mV for an 0 to 200 mmHg range) generated by the applied pressure is preamplified by a bridge-type amplifier (Figure 8). A second amplifier provides additional gain and offset capability. The signal is recorded on magnetic tape and displayed on a standard strip chart recorder. The derivative of the pressure signal is obtained from an operational amplifier circuit.

Electronic calibration of the overall pressure-measuring system, exclusive of the transducer, at the start and end of an experiment is achieved by substituting a precision resistive divider network for the pressure probe (Figure 9). Bridge imbalance, as would be caused by an applied pressure, is thereby simulated by changing the values of the resistors in the calibration network and monitoring the offset voltage produced. These reference levels are later converted to actual millimeters of mercury by referring to the most recent in vitro or in vivo calibration of the transducer.

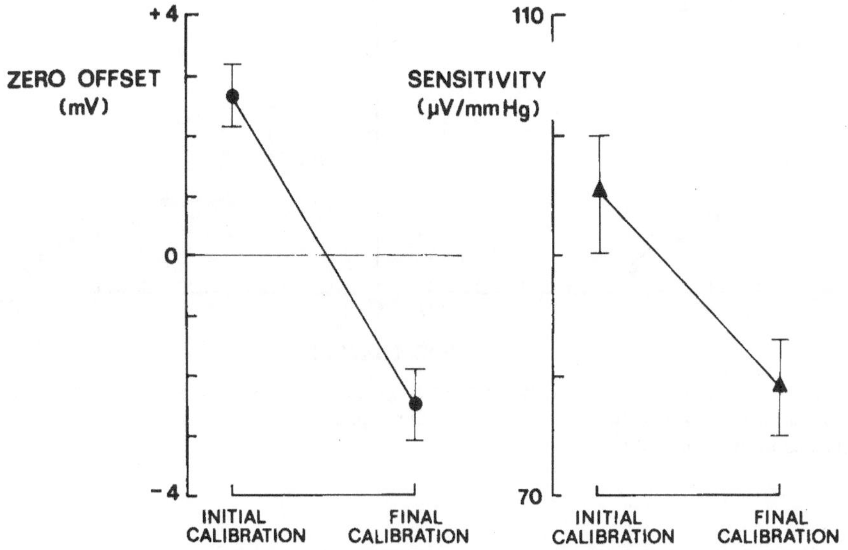


FIGURE 4. Values (mean ± SEM) of zero pressure offset and sensitivity determined from an in vitro calibration immediately following the first chronic implant, and from a subsequent in vitro calibration after the gauges had been regularly used for 80 weeks.

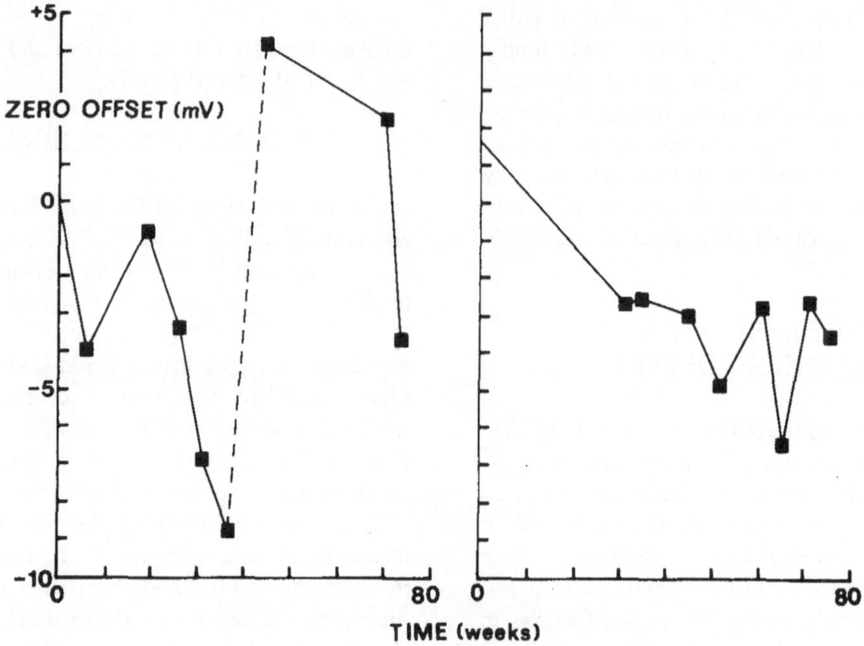


FIGURE 5. Typical variations in zero pressure offset for two gauges over a period of 80 weeks. Note the irregular changes with time, although there is a general tendency for the offset to fall. The broken line represents the point when the gauge was returned to the manufacturer.

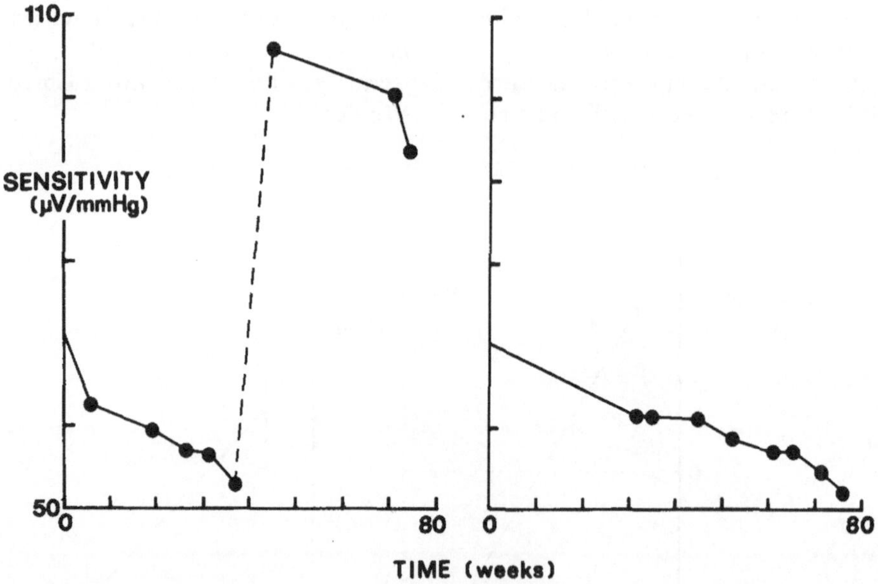


FIGURE 6. Variations in sensitivity for the two gauges in Figure 5, showing the continuous decrease in sensitivity with time. The broken line represents the point when the gauge was returned to the manufacturer.

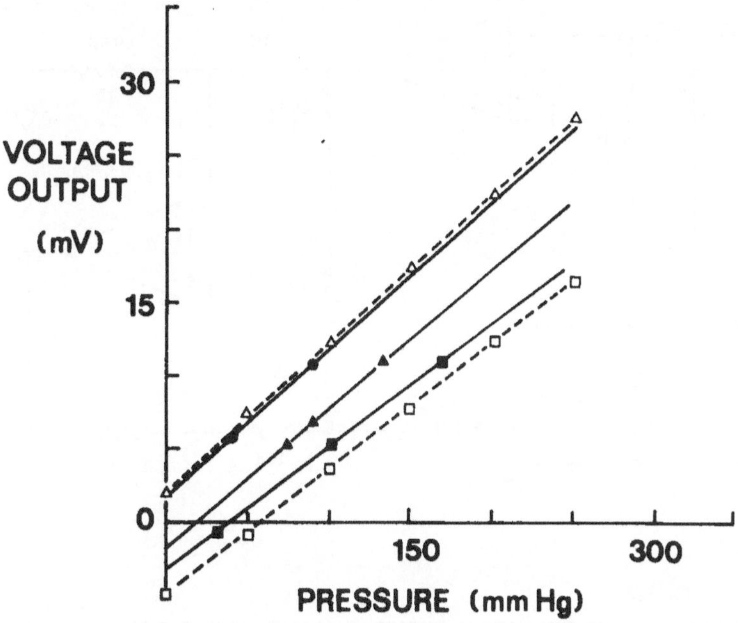


FIGURE 7. Repeated calibrations, both in vivo (dashed line) and in vitro (solid line), for a single gauge over a 12-week period. While the zero offset is clearly reduced, there is little change in the slopes of the lines (sensitivity).

Simultaneous comparison of the output of the implanted pressure gauge and a calibrated microtip manometer pressure transducer or catheter and strain gauge manometer is the most accurate method of calibration. The animal's blood pressure is varied transiently with an appropriate pressor agent to achieve a wide range of calibration points for comparison. This procedure overcomes the problem of changes that may have occurred in the solid state gauge's transfer function since the last in vitro calibration. Figure 7 illustrates the in vivo vs. in vitro calibrations for one gauge over a 3-month period and demonstrates that the drift of the transducers may be followed using these techniques. In our experience, this procedure must be repeated approximately every 2 weeks for aortic implants, but daily cross-calibration is necessary for experiments in which zero pressure level is critical, e.g., measurement of end-diastolic or atrial pressure. For these experiments requiring daily calibration, it is advisable to implant a catheter in the ventricle or atrium.

## CONCLUSION

While many of the problems associated with the measurement of pressure in animals with long-term chronic implants have been discussed above, it is clear that the inherent advantages of the implantable gauges make them, in principle, very attractive. Moreover, as previously pointed out, there are many situations where they are the only type of pressure transducer that may be used. The major limitations that must be overcome are drift in zero pressure offset and sensitivity, and lack of electronic calibration reference to atmospheric pressure. Finally, reduction in size of the transducer and cable would allow application to experiments involving use in smaller animals, e.g., smaller primate studies and fetal preparations.

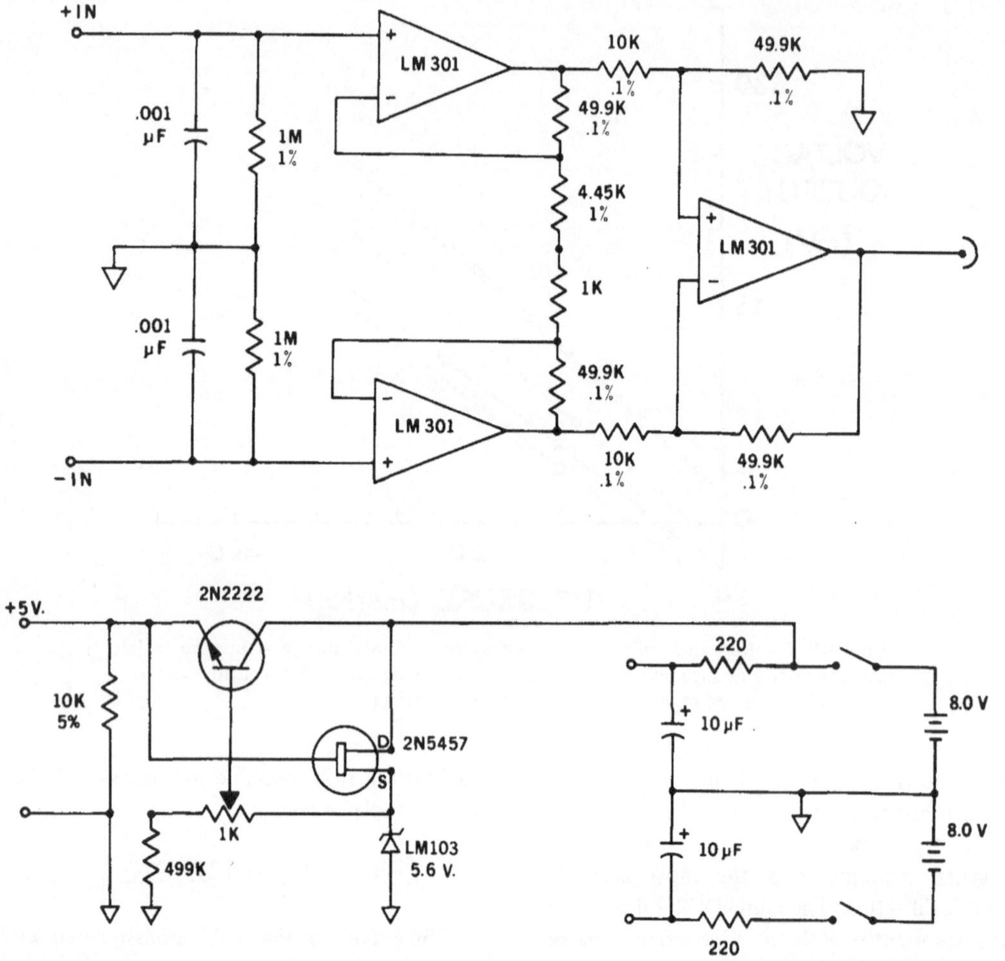


FIGURE 8. Instrumentation amplifier used to amplify the pressure signal generated by a solid state gauge. The ± 8 V used to power the operational amplifiers and 5 V excitation for the pressure gauge are provided by the power supply shown in the bottom of the figure.

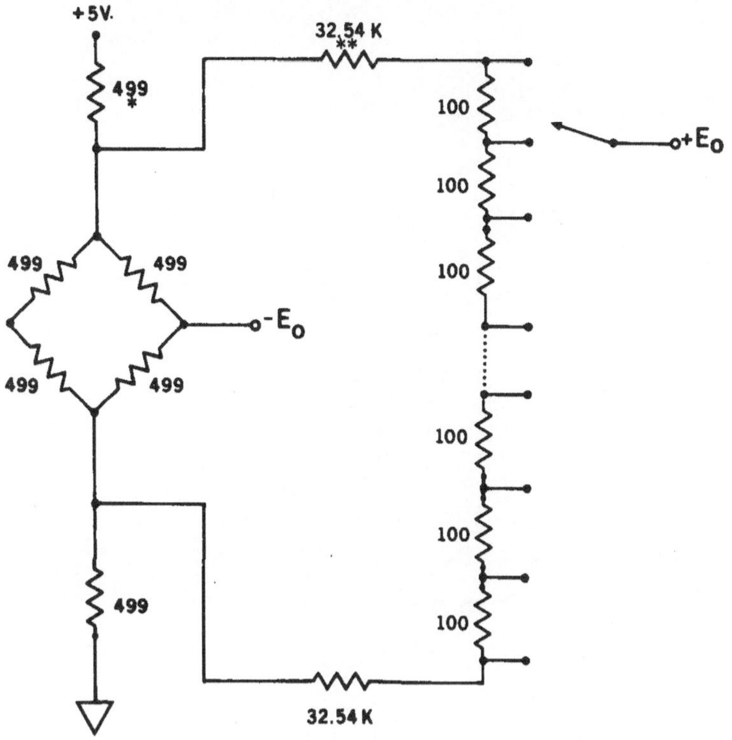


FIGURE 9. The Wheatstone bridge of the pressure gauge is simulated by this precision resistive divider network. There are a total of sixteen 100-Ω resistors which will allow voltage offsets to be selected in 2.5-mV steps for the electronic calibration of the pressure-measuring system. All resistors are 0.1%. The 499-Ω resistor marked with an asterisk and the 32.5-kΩ resistor marked with two asterisks should be trimmed to balance the bridge to obtain the proper zero and full scale readings.

## REFERENCES

1. **Van Citters, R. L. and Franklin, D. L.**, Telemetry of blood pressure in free-ranging animals via an intravascular gauge, *J. Appl. Physiol.*, 21, 1633, 1966.
2. **Vatner, S. F., Franklin, D. L., Higgins, C. B., Patrick, T., and Braunwald, E.**, Left ventricular response to severe exertion in untethered dogs, *J. Clin. Invest.*, 51, 3052, 1972.
3. **Vatner, S. F., Higgins, C. B., White, S., Patrick, T., and Franklin, D.**, The peripheral vascular response to severe exercise in untethered dogs before and after complete heart block, *J. Clin. Invest.*, 50, 1950, 1971.
4. **Vatner, S. F., Patrick, T. A., Higgins, C. B., and Franklin, D.**, Eating and digestion in conscious unrestrained primates, *J. Appl. Physiol.*, 36, 524, 1974.
5. **Van Citters, R. L., Franklin, D. L., Vatner, S. F., Patrick, T., and Warren, J. V.**, Cerebral hemodynamics in the giraffe, *Trans. Assoc. Am. Physicians*, 82, 293, 1969.

# *Present Applications of Indwelling Pressure Transducers*

# SURVEY OF PRESENT APPLICATIONS FOR INDWELLING PRESSURE TRANSDUCER SYSTEMS, AND EXPERIENCES WITH CARDIOVASCULAR IMPLANTS

**Ernest P. McCutcheon**

In theory, pressure measurement is important for understanding function in every part of the body. In practice, problems of transducer size and biological tolerance confine major applications of implanted pressure measurement to hollow organs such as the bladder and heart; areas of potential or artificially created spaces such as intrapleural, peritoneal, and penile sites; and the interior of implanted devices such as perforated cylinders or prostheses. The continued miniaturization of transducers provides an increasing number of options for experimental and clinical application, with particular impact on the long-term placement of instrumentation. In our laboratory, intraventricular and intra-aortic pressure transducers have been tolerated in dogs for periods exceeding 7 years. Changes at necropsy in five dogs sacrificed after implantation periods ranging from 1 to 7.8 years included localized fibrosis at the implantation site, with occasional but significant evidence of embolic phenomena. An additional five dogs are asymptomatic after 1 to 5 years. These findings indicate a need for improvement not only in size, but biocompatibility, especially at the point of fixation and the interface between the transducer and tissue. A further requirement is improved zero stability and accuracy to measure relatively small pressure differences.

## INTRODUCTION

A broad class of functional activity is governed by the absolute level of force distributed over an area (or pressure) and pressure differences or gradients. These processes range from macroscopic events such as cardiac contraction, respiration, digestion, and urination, to microscopic events such as exchange of gases and fluids across cell membranes. Pressure is of such widespread importance in biological processes that the basic measurement is very familiar. In practice, difficulties with access to the site of the desired pressure measurement often require many compromises, and numerous methods have been devised to improve the range of available techniques. A fluid-filled catheter connection from the measurement site to the transducer remains a standard approach, but the catheter transmission line carries many problems, including susceptibility to mechanical artefacts, limited frequency response, and loss of continuity or patency caused by factors such as clotting or compression. The drawbacks are particularly marked in animals with high heart rates and small blood vessels, and they are especially severe in research and monitoring situations requiring long-term recordings (weeks, months, or years).

Many of the difficulties with catheters have been overcome by the advent of miniaturized transducers which can be placed directly at the measurement site. Continued development of special construction techniques has shown the feasibility of small units based on resistive, inductive, capacitive, and fiber-optic principles,[1] but the availability of semiconductors allowed a breakthrough in miniaturization and manufacturing economy.[1-3] The semiconductor class of materials has high gauge and resistivity factors; the most widely used representative of this class is silicon. This versatile material may be used in many forms: bonded, unbonded; through anisotropic stress effects; with resistors diffused on a silicon chip; or by deposition directly (thin-film) on the force-orienting diaphragm. A number of these devices are available commercially in 3 to 6.5-mm diameters.[3] Recent technological advances, derived from integrated circuit processes and providing further size reduction, are also discussed in this text (see chapters by W. H. Ko, T. A. Nunn and J. B. Angell, and J. Hynecek).

Within the constraints of size and biological tolerance, three major application groups can be identified. The first group includes sites or organs with an accessible space, either actual or potential. Suitable actual spaces include the ear, mouth, nasal passages, throat, and trachea as representative examples from the upper airway. The esophagus and other gut segments, including the gall bladder, offer available cavities within the gastrointestinal tract. In the genitourinary system, the urinary tract, bladder, and uterus are accessible. Other locations more nearly potential in nature include cerebrospinal, intrapleural, peritoneal, joint, penile, and intramuscular (skeletal and cardiac) areas.

A second major group is the space created artificially with devices such as an implanted wire mesh capsule or fibrous wick. These techniques are usable in almost any solid tissue.

The third major group is that of prosthetic devices placed for correction of functional pathology; transducers may be included for management of the implant or to gain information on function not otherwise available.

Therefore, extensive applications of implantable transducers exist. The requirements range from analysis of organ function, including mechanical properties under static and dynamic loads, to system regulation. Immediate objectives range from acquisition of primary research data in humans and restrained or unrestrained animals, to diagnostic, surgical, and intensive care monitoring in patients. A more distant but rapidly developing application is the long-term modification and control of a number of physiological processes, through implanted devices replacing damaged parts or infusing biochemically active agents, or by means of biofeedback. The subject environment may be normal or hostile, ranging from underwater to spacecraft. These considerations indicate the broad significance of improving the state-of-the-art for pressure measurement using indwelling pressure transducers.

## REPRESENTATIVE APPLICATIONS

The following sections include highlights of the uses of indwelling pressure transducers, concluding with a summary of our experiences with the placement of these devices in the cardiovascular system of animals.

Virtually every approach noted in the introduction has been taken in various basic and clinical investigative programs, both animal and human. For the application group of organs with an accessible space, beginning with the region of the upper airway, force exerted by the lip against an orthodontic appliance has been recorded for prolonged periods.[4] The signal was obtained from a strain gauge directly wired to a belt-carried tape recorder system. Force levels were encoded into a pulse train for data compression and magnetic tape storage. Using this system, continuous data were acquired for periods as long as 8 hr. At a related site within the ear, the levels of middle ear pressure are significant for analysis of the hearing process; negative pressure in the middle ear can cause the appearance of a transudate and conductive hearing loss within 5 to 15 min.[5] Barga and Miller[5] have begun development of a rhesus monkey preparation for study of the Eustachian tube and middle ear pressure-function relationship. In the initial preparation, only a chronically implanted catheter was used. Postoperative effusions obstructed the cannula in two of the four monkeys. Later, the obstructions tended to clear, but all the catheters were irreversibly occluded after approximately 3 months. A chronically implanted pressure transducer was not used, but would appear highly applicable.

For study of gastrointestinal motility and transport, swallowable pressure transducers with direct connections, or in capsules with circuits for telemetering signals, are well known.[6] A discussion of such gastrointestinal applications is included in this text (see chapter by A. Connell).

Utilization for analysis of urinary bladder regulation is extensive. Timm and co-workers reported on their experiences with a telemetry capsule for monitoring cuff pressure in an artificial urethral sphincter and pointed out the susceptibility of the transmitter to temperature variation.[7,8] Our experience with noncrystal controlled telemetry systems is similar.

Impressive applications for intrauterine monitoring of fetal heart rate and intrauterine pressure in humans are underway. In one study, data were obtained from ten patients for a mean duration of 215 min.[9] Silicone-covered piezoresistive transducers placed in various positions within the uterus showed that no single point provides a representative pressure.[10] Accurate measurement of intrauterine pressure requires integration of the output from multiple transducers. In this context, the need to distinguish between mechanical impact forces and pressure has been emphasized by Neuman and colleagues.[10] The stainless steel shroud they used to avoid the impact problem may be pertinent in other applications. For instance, from the group of locations with potential spaces, a pressure transducer was implanted in the corpus cavernosum of the bull penis to study the cause of penile hematoma.[11,12] During copulation, very brief, rhythmic transients were recorded, with a mean peak value of approximately 14,000 mmHg, and maximum value slightly exceeding 32,000 mmHg. It is difficult to accept such levels as representing the true state of the spatial hydrostatic pressure within the penis. A direct mechanical force seems much more probable.[13]

Varied approaches have been taken to solve the problem of recording intracranial pressure.[14] An indwelling transducer concept tested in animals provided much less artefact and better sealing than open-ended or balloon-tipped catheters connected to external transducers.[15] Recent developments for this application are discussed in this text (see chapter by A. E. Walker).

The significance for wound healing of the interaction between intra-abdominal pressure and wall forces has been demonstrated,[16] as well as the relationship of abdominal pressure and spinal strain.[17]

The potential for obtaining data from humans appears to have been scarcely tapped, especially when artificial and prosthetic devices are implanted. Rods used for correcting scoliosis have been instrumented with inductively coupled units transmitting strain data.[18] Carlson and associates described a device for measuring the magnitude and distribution of pressure forces generated between the surface of the human hip socket and a hip prosthesis replacing the femoral head.[19,20] An array of 14 transducers is positioned within the prosthesis, and the signal is telemetered from transcutaneous inductively powered equipment located within the replacement femoral head.

The use of artificial devices to control urinary bladder function was discussed earlier.[7,8] Various cardiovascular prostheses, such as valve rings, also lend themselves to similar purposes (see chapter by R. J. Kiraly, G. Jacobs, and Y. Nosé).

## ANIMAL CARDIOVASCULAR IMPLANTS

Indwelling pressure transducers have been used extensively for evaluation of cardiovascular function in chronically implanted animals. For over 10 years, we have been placing pressure transducers in the heart and major vessels, with both directly wired and multichannel telemetry types of signal acquisition. In over 200 dogs with percutaneous leads from implanted cardiac and arterial pressure transducers, the most common complications caused by the transducers have been embolic phenomena and, occasionally, extrusion from the intravascular site. The animals with telemetry invariably have a much smoother postoperative course and appear to recover from surgery more rapidly than those with percutaneous leads. From clinical criteria, they tolerate totally implanted devices for a longer period with less difficulty. Long-term tolerance and operating life assume added significance in view of the recent finding from our laboratory that dogs chronically implanted with transducers connected to percutaneous leads require at least 3 weeks recovery before cardiovascular responses stabilize.[21]

Because of the greater tolerance and ease of care, our major long-term experience is with implantations of transducers incorporated into telemetry units. In a previous study, Sandler et al.[22] reported experiences in 17 dogs and 8 chimpanzees. Representative results are those from post-mortem examinations of 3 of the 17 dogs sacrificed 1 month after implantation of a multichannel telemetry unit with 6.5-mm diameter pressure transducers in the ventricle and aorta.[22] No embolic phenomena from the intravascular transducers were identified, and fibrous reaction to the unit was minimal. In these and other dogs discussed below, the transducer surface was prepared for implantation by applying a coat of heparinized colloidal graphite in alkyd resin (Dag® 35, Acheson Colloids Co., Port Huron, Michigan).

Ten additional dogs have been implanted for periods exceeding 1 year. Recently, five of them were sacrificed after implantation periods ranging from 1 to 7.8 years. In four of the animals (4 to 7.8 years), the pressure transducers (6.5-mm diameter, Konigsberg Instruments) were placed in the left and right ventricles and descending thoracic aorta, and a multichannel battery-powered telemetry unit was positioned in the intrapleural space. In the fifth animal (1 year), there were two pressure transducers (left ventricle and descending thoracic aorta) and an electromagnetic (EM) flow transducer on the ascending aorta. In the one animal necropsied after 7.8 years, the transducers were well tolerated without signs of embolism (Figure 1). The batteries in the telemetry unit out-gassed and disintegrated, and the lead wires were also deteriorated markedly; however, these changes were contained by the usual local fibrotic reaction. Tolerance to the

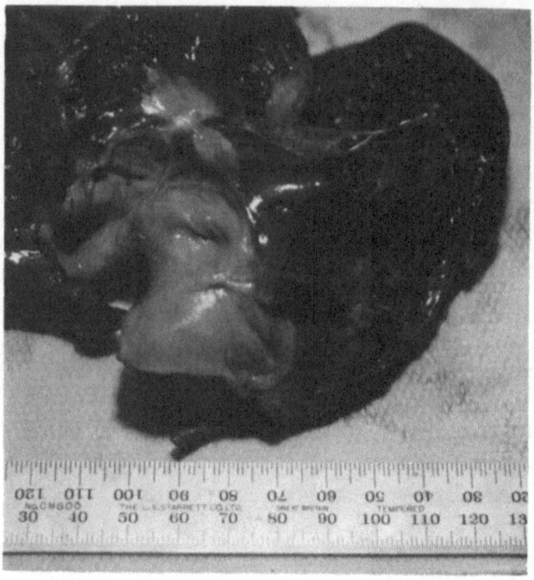

FIGURE 1. The appearance of a right ventricular pressure transducer 7.8 years after implantation. The portion of the transducer within the heart is completely covered by a smooth layer of endothelium (black arrow).

silicone-coated telemetry units is consistently high, with only localized fibrosis typically present. Two other animals also showed no unexpected gross alterations.

In the fourth animal (5.6 years), also with disintegrated batteries, extensive changes had occurred. Inflammation was present at the battery site, and a lung abscess and renal infarcts were found. The fifth animal died from rupture of the aorta at the EM transducer site. At post-mortem, a recent clot was attached to the pressure transducer in the descending aorta. The clot appeared to be an agonal occurrence. The left ventricular transducer was endothelialized and appeared normal. In most cases the Dag layer had disintegrated and disappeared; the transducers were otherwise intact. The remainder of the morphological changes in all dogs were those expected following thoracotomy. The other five dogs remain clinically normal 1 to 5 years after implantation. The long-term tolerance of dogs to chronically implanted, intravascular pressure transducers appears reasonably satisfactory, though not a guaranteed outcome.

In a related study with similar results by Zech et al., seven rhesus monkeys held in chair restraint were chronically implanted with a left ventricular pressure transducer connected by exteriorized leads.[23] Two of the monkeys died 1 month after surgery; the other five were sacrificed 3 to 5 months after implantation. In six of the seven, necropsy showed chronic adhesions of the visceral and parietal pleura (typical post-thoracotomy changes), with some chronic localized epicarditis and focal myocardial degeneration at the transducer insertion site. Acute multifocal renal infarcts suggestive of emboli were evident in three of the seven animals. They concluded there was a relationship between the size of the transducer placed in the heart and the occurrence of emboli. Transducer diameters were 6.2 and 3 mm. The larger transducer was used in two of the three animals with emboli. No statement was made regarding efforts to provide an antithrombogenic coating over the transducer face.

Because of recurrent problems with battery failure and excessive size of multichannel telemetry units, we are evaluating an inductively powered model small enough for monkeys (Figure 2). This is the unit discussed in the chapter by H. L. Stone. Our initial results in macaques and small dogs (9 to 11 kg) indicate satisfactory tolerance, with no apparent complications from 3 weeks to 1 year after surgery in five monkeys and three dogs.

## PROBLEMS AND PRECAUTIONS

Some of the potential sources of difficulty were alluded to in the preceding discussion, but more emphasis is deserved. The essential areas of concern can be summarized as: (a) biological compatibility, (b) placement technique, and (c) measurement accuracy. Even without systemic alterations occurring for other reasons, such as infection, the incidence of embolism is significant and troublesome. Much more attention must be paid to reactions at the transducer interface. For instance, antithrombogenic coatings still need improvement; the incidence of embolism ranges from 3 to 40%.[22,23] We now routinely use the tridodecylmethyl ammonium chloride-heparin solution (TDMAC-heparin complex, Polysciences, Inc., Warrington, Pennsylvania) in our laboratory. This material appears to provide a more effective antithrombogenic action than Dag 35-heparin. Clearly, either agent is better than nothing. The thin-film silicon transducers often have a silicon dioxide surface. Since glass is highly thrombogenic, this layer may cause problems, unless high purity confers protection. A recent report suggests that surface bonding of a fibrinolytic enzyme improves biocompatibility of foreign

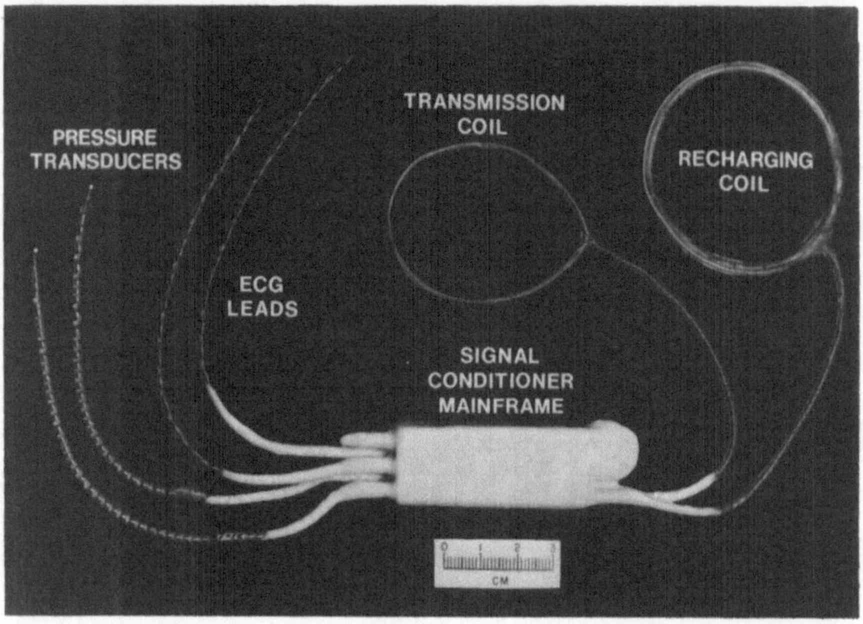


FIGURE 2. A multichannel telemetry system incorporating indwelling pressure transducers. The entire system is suitable for implantation within the thorax of monkeys or dogs as small as 9 to 10 kg. The unit is powered by inductive coupling from a coil placed on the skin over the implanted coil.

materials.[24] This and other approaches should be thoroughly evaluated.

The local inflammatory response may also be a significant factor. The methods of insertion and fixation at the vascular entry site are far from ideal; a high level of technical skill is necessary to perform this step successfully. Sutures, the mainstay of the transducer-anchoring process, are relatively crude and cumbersome, undoubtedly enlarge the area of inflammation, delay healing, and increase scarring. The bucrylate monomer adhesives were promising alternatives to sutures, but are no longer manufactured. Transducer designers should aim for configurations which simplify placement.

In our experience, the contribution of the signal-processing system to measurement error has been minor. The unavoidable characteristics of the bonded semiconductor strain gauges have been the more overriding concern.[3] The transducers are sealed, and the zero reference will vary with barometric pressure. Temperature-dependent instabilities may also be present. Careful repeated calibration, both static and dynamic, is essential prior to implantation, preferably at the temperature of the implantation site (38°C in dogs). In our laboratory, this process is aided by a temperature-controlled water bath with adapters for transducers and leads of various sizes (Figure 3). The insert chamber is filled with Detergicide® or a similar solution to maintain sterility during testing. For dynamic response evaluation, a loudspeaker-cylinder system (supplied by Millar Instrument Co., Houston, Texas) is durable and reasonably priced.[25]

A positive benefit of the sealed transducer is testing of gain stability after implant by placing the animal in an altitude chamber and introducing known pressure steps, as described in the chapter by H. L. Stone.[22] In the final analysis, however, the best way to know what is happening is by periodic comparison to a suitable reference standard, preferably a manometer-tipped catheter such as those available from Millar Instrument Co.[26] Alternatively, a fluid-filled catheter-manometer system can be used, but it must be well characterized.[27]

A final caution should be noted for applications where direct mechanical impact may be present, as in the uterus, a joint, a small-volume left ventricle, and related potential spaces. The presence of unusually high transient pressure levels should raise suspicion of direct impact as the source.

An accurate catheter-manometer pressure

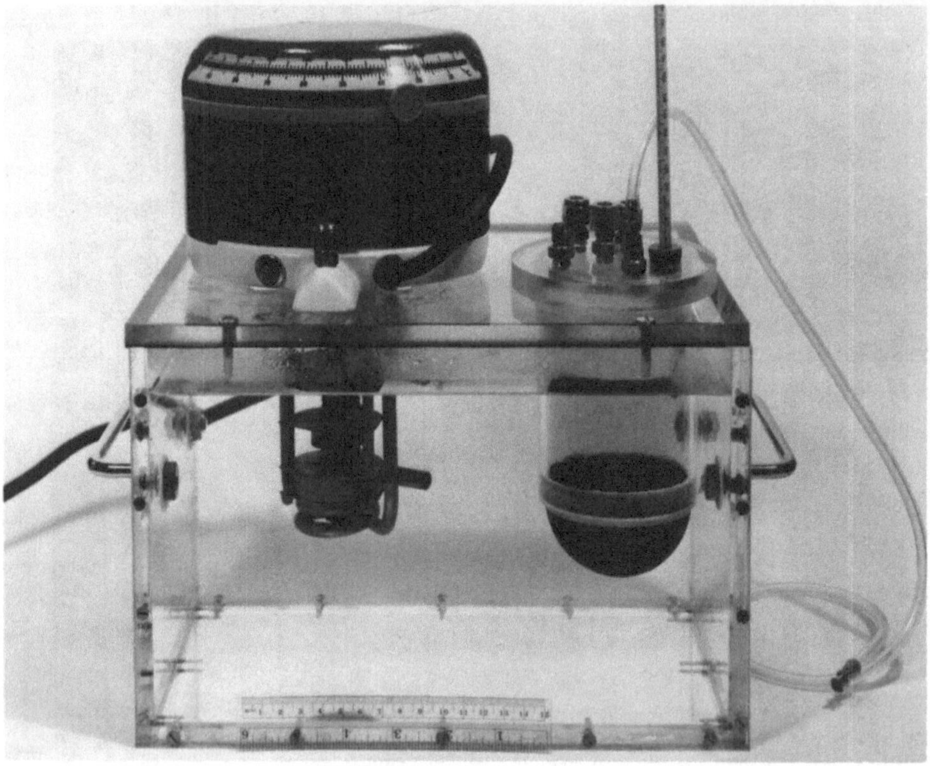

FIGURE 3. A temperature-controlled water bath for static calibration of indwelling pressure transducers or manometer-tipped catheters of various dimensions. A bactericidal solution in the removable cylindrical insert maintains sterility during the testing phase.

measurement cannot be obtained from an animal while it is running or swinging from a limb. The availability of indwelling pressure transducers has greatly widened the spectrum of useful techniques for research and clinical applications. When combined with telemetry, it is possible to study subjects whether restrained or free-ranging, on the ground, underwater, or at altitude and weightless.

## SUMMARY

Pressure measurement is important for understanding function in every part of the body. At the microscopic level, pressure gradients govern exchange of materials such as water and gases. At the macroscopic level, pressure and especially pressure differences are significant in every organ. Major applications of implanted direct pressure measurement include: (a) hollow organs such as the bladder, gut, heart, airway, and uterus, and areas of potential or artificially created spaces such as intrapleural, peritoneal, penile, and cerebrospinal sites; (b) the interior of implanted devices such as perforated cylinders; and (c) components of prosthetic implants.

The continued miniaturization of transducers provides an increasing number of options for experimental and clinical applications, with particular impact on the long-term placement of instrumentation. For example, ten dogs with intraventricular and intra-aortic pressure transducers have tolerated the presence of these devices for periods exceeding 7 years. Changes at necropsy in five dogs sacrificed after implantation periods ranging from 1 to 7.8 years included localized fibrosis at the implantation site, with occasional evidence of embolic phenomena, primarily recent and healed renal infarcts. The other five dogs are asymptomatic after 1 to 5 years. These findings indicate a broad range of applications for such devices, but improvements must be made in biocompatibility and placement technique, especially at: (a) the point of fixation and (b) the interface between the transducer and tissue. Further needs are adequate zero stability for the accuracy required to measure relatively small pressure differences in locations such as the intrapleural space and the interior of solid organs, and smaller size to extend the range of applications.

# REFERENCES


1. **Cobbold, R. S. C.,** Transducers for biomedical measurements: Principles and applications, in *Wiley-Interscience Series, Biomedical Engineering and Health Systems,* Milsum, J. H., Adv. Ed., New York, John Wiley & Sons, Inc., 1974, 190.
2. **Angelakos, E. T.,** Semiconductor pressure microtransducers for measuring velocity and acceleration of intraventricular pressures, *Am. J. Med. Electron.,* 3(4), 266, 1964.
3. **Konigsberg, E.,** A pressure transducer for chronic intravascular implantation, *Biomed. Sci. Instrum.,* 3, 259, 1967.
4. **Beeler, G. E. and Bleil, B. T.,** Long-term free-ranging recording using data integral to pulse frequency signal transformation, *IEEE Trans. Biomed. Eng.,* 21, 245, 1974.
5. **Barga, J. L. and Miller, J. M.,** Eustachian tube function in the rhesus monkey, *Arch. Otolaryngol.,* 99(2), 105, 1975.
6. **Staf, J. J., Dodds, W. J., Hogan, W. J., Linehan, J. H., and Stewart, E. T.,** Intraluminal esophageal manometry: An analysis of variables affecting recording fidelity of peristaltic pressures, *Gastroenterology,* 67(2), 221, 1974.
7. **Timm, G. W., List, J. S., Bradley, W. E., and Scott, F. B.,** Frequency stability of an implantable pressure telemetry capsule, *IEEE Trans. Biomed. Eng.,* 22(3), 250, 1975.
8. **Timm, G. W., Bradley, W. E., and Scott, F. B.,** Experimental evaluation of an implantable externally controllable urinary sphincter, *Invest. Urol.,* 11, 326, 1974.
9. **Neuman, M. R., Picconatto, J., and Roux, J. F.,** A wireless radiotelemetry system for monitoring fetal heart rate and intrauterine pressure during labor and delivery, *Gynecol. Invest.,* 1, 92, 1970.
10. **Neuman, M. R., Jordan, J. A., Roux, J. F., and Knoke, J. D.,** Validity of intrauterine pressure measurements with transcervical intra-amniotic catheters and an intra-amniotic miniature pressure transducer during labor, *Gynecol. Invest.,* 3(5), 165, 1972.
11. **Beckett, S. D., Walker, D. F., Hudson, R. S., Reynolds, T. M., and Vachon, R. I.,** Corpus cavernosum penis pressure and penile muscle activity in the bull during coitus, *Am. J. Vet. Res.,* 35(6), 761, 1974.
12. **Beckett, S. D., Reynolds, T. M., Walker, D. F., Hudson, R. S., and Purohit, R. C.,** Experimentally induced rupture of corpus cavernosum penis of the bull, *Am. J. Vet. Res.,* 35(6), 765, 1974.
13. **Shanks, C. and Harrison, G. A.,** Mechanical factors affecting the evoked pressures from a simple twitch monitor, *Anesth. Intensive Care,* 1(2), 154, 1972.
14. **Zattoni, J., Siani, C., and Semino, L.,** Some technical factors affecting the intracranial pressure recorded by epidural space-occupying devices, *Acta Neurochir.,* 24(4), 353, 1973.
15. **Jacobson, S. A. and Rothballer, A. B.,** Prolonged measurement of intracranial pressure using an absolute pressure transducer, *Biomed. Sci. Instrum.,* 4, 179, 1968.
16. **Williams, J. F., Kirpatrick, J., and Syme, G. A.,** Force measurement in the abdominal wall, *BioMed. Eng.,* 10(5), 175, 1975.
17. **Eie, N.,** Recent measurements of the intra-abdominal pressure, in *Perspectives in Biomedical Engineering,* Kenedi, R. M., Ed., University Park Press, Baltimore, 1973, 121.
18. **Nachemson, A. and Elfstrom, G.,** Intravital measurement of forces in the human spine: Their clinical implications for low back pain and scoliosis, in *Perspectives in Biomedical Engineering,* Kenedi, R. M., Ed., University Park Press, Baltimore, 1973, 111.
19. **Carlson, C. E., Mann, R. W., and Harris, W. H.,** A look at the prosthesis-cartilage interface: Design of a hip prosthesis containing pressure transducers, *J. Biomed. Mater. Res.,* 8(4), Part 2, 261, 1974.
20. **Carlson, C. E., Mann, R. W., and Harris, W. H.,** A radio telemetry device for monitoring cartilage surface pressures in the human hip, *IEEE Trans. Biomed. Eng.,* 21(4), 257, 1974.
21. **Carlson, E. L., McCutcheon, E. P., Gordon, R. M., and Sandler, H.,** Postoperative dose-response changes to isoproterenol and norepinephrine in chronically instrumented dogs, *Pharmacologist,* 16(2), 446, 1974.
22. **Sandler, H., Stone, H. L., Fryer, T. B., and Westbrook, R. M.,** Use of implantable telemetry systems for study of cardiovascular phenomena, *Circ. Res.,* 30–31, Suppl. II, 85, 1972.
23. **Zech, W. H., Boucher, J. H., Hilmas, D. E., and Spertzel, R. O.,** Myocardial contractility in rhesus monkeys: A method for long-term study, *Am. J. Vet. Res.,* 35(1), 83, 1974.
24. **Nguyen, A. and Wilkes, G. L.,** A thromboresistant surface by enzyme immobilization, *J. Biomed. Mater. Res.,* 8, 261, 1974.
25. **McCutcheon, E. P., Evans, J. M., and Stanifer, R. R.,** Evaluation of miniature pressure transducers, in *Chronically Implanted Cardiovascular Instrumentation,* McCutcheon, E. P., Ed., Academic Press, New York, 1973, 21.
26. **Millar, H. D. and Baker, L. E.,** A stable ultraminiature catheter-tip pressure transducer, *Med. Biol. Eng.,* 11, 86, 1973.
27. **Asmussen, M., Lindstrum, K., and Ulmsten, U.,** A catheter-manometer calibrator – a new clinical instrument, *BioMed. Eng.,* 10(5), 175, 1975.

# STUDIES OF INTESTINAL MOTILITY

**Alastair Connell**

The study of intestinal motility has traditionally been confined to the study of pressures within the lumen of the bowel or transmitted through a balloon transducer in contact with the wall of the bowel. Analogue recordings have been interpreted as giving information about the contractions of the smooth muscle of the bowel wall. In addition, studies of intestinal transit have implied a variety of methods to determine the movement along the bowel of appropriate markers. The tone of the degree of rigidity of the bowel wall has only been very imprecisely measured.

Underlying the mechanical contractions of the bowel wall are the electrical rhythms. Much less is known about these in man, although animal studies are well developed. Methods to record the electrical rhythms in man are rudimentary and few reliable data exist.

The terminal ileum and the distal colon are relatively inaccessible and radiotelemetering has been used to derive information from these areas. The untared radiotelemetering capsule is, however, somewhat restricted in that it cannot be retained in one place and inevitably moves in response to gut motor activity. However, by utilizing the directional nature of the strength of the signal from a telemetering capsule, some information about these movements can be obtained at the same time as the pressure record.

Gastrointestinal motor disorders underlie and are associated with many common complaints, and only rudimentary studies have been made to understand the nature of these disorders. Advances in technology would greatly improve the penetration of our diagnostic and possibly therapeutic efforts.

A general overview of motility recordings from the gastrointestinal tract should be included at this point. The term motility is one of the less precise usages of physiology and medicine, and some confusion arises because of this imprecision. Motility can mean anything from watching or measuring the contractions of muscle strips in vitro to various parameters in the intact person. We are concerned with essentially three different features in man: (1) the transit of material along the alimentary tract; (2) the contractions of the muscle of the alimentary tract, which can be examined by radiology or more directly by measurement of intraluminal pressures, and the electrical rhythms of the alimentary tract; and (3) the estimation of the "tone" of the alimentary tract. "Tone" is relatively easy to conceptualize but difficult to measure. The nearest approach we get to measurement of this particular function is to determine pressure-volume curves by some form of balloon system. No one has seriously attempted the study of tone in the intact man.

Both intraluminal and external transducers have been used for the study of alimentary motility. Most clinicians and many researchers have used external transducers with fluid-filled catheters. By and large these are satisfactory. The pressure changes in the alimentary tract are slow, and the response of a transducer and catheter assembly is completely adequate to capture even the fastest components of these pressure changes. Intraluminal transducers of the Gauer-Gienapp type have also been applied. More recently, telemetry devices have been employed and these will be discussed shortly.

The areas of the GI tract which have been studied are those which are easily accessible to the catheters, so that not surpisingly the esophagus and the fundus of the stomach have been very thoroughly explored. The rectum and the sigmoid colon have been relatively well explored, but the long tract in between, with the exception possibly of the duodenum, is almost virgin territory. The reason for this is quite simple. It may take 2 or 3 days to have the catheter passed to the terminal ileum or the cecum, and if a volunteer does that once, he is seldom enthusiastic about the second or third occasion. This is a very real limitation on experimentation. Telemetering has helped a little, but is not free from problems.

Another area which is very little explored in man is the recording of electrical rhythms from the gut. We have fairly extensive knowledge of the nature of these rhythms in animals, and we have a degree of understanding of the relationship between the potentials and muscle contractions. But in man there has been a real inhibition about employing these methods, because of the difficulty of the technology of insuring that the potentials detected, in fact, reflect the potentials generated by the muscle. Intraluminal sensors are suspected in that there are other potentials which may be detected by the intraluminal sensor, and there is concern about the faithfulness of the muscle recording. Figure 1 illustrates a record taken from the alimentary tract in man, obtained using an intraluminal catheter associated with a small, metal tambour. Movement of the tambour deviated a light beam and the analogues recorded on photographic paper. The records which are obtained indicate that we have a complex type of analog in most areas of the alimentary tract.

Progress has been delayed because of inadequate attention to analysis. Attempts at pattern recognition or analog analysis have, for the most part, been sporadic. Clinicians are still inspecting these records visually and obtain no more than gross overall assessments of activity. Little effort has been made to apply more sophisticated analytical techniques to the studies, and we need to focus more on interpretation of recording rather than simply on generation of the data.

If we are to have information about the more inaccessible areas of the alimentary tract, we need a device which is not tethered. Many groups have explored the use of telemetering capsules and have, to an extent, derived information from the small intestine and the cecum. An early version was developed in the late 1950s and early 1960s.[1,2] For several reasons, there has been little further development of this device. It has a somewhat large size, measuring 2.2 cm × 0.8 cm. There is no problem swallowing it, but there are some problems which might alter its movement along the alimentary tract. The pylorus may not readily pass objects of this size and shuttles them back and forth for as many as 12 to 18 hr before passing them into the duodenum. This is extremely inconvenient for the clinical investigator and his patient because for some reason the small intestine seems to be much more receptive between midnight and 4:00 a.m. than at any other time. That upsets patients, physicians, and particularly nurses. Thus, size is quite critical and a second generation of capsules which are of small size is needed. With newer technology the size factor may be improved. The pressure transducer in these capsules is a diaphragm associated with the lid of a pot case. The coil alters the frequency of the oscillating circuit. An external antenna detects the signal which is displayed in the usual ways. Telemetering capsules increase the

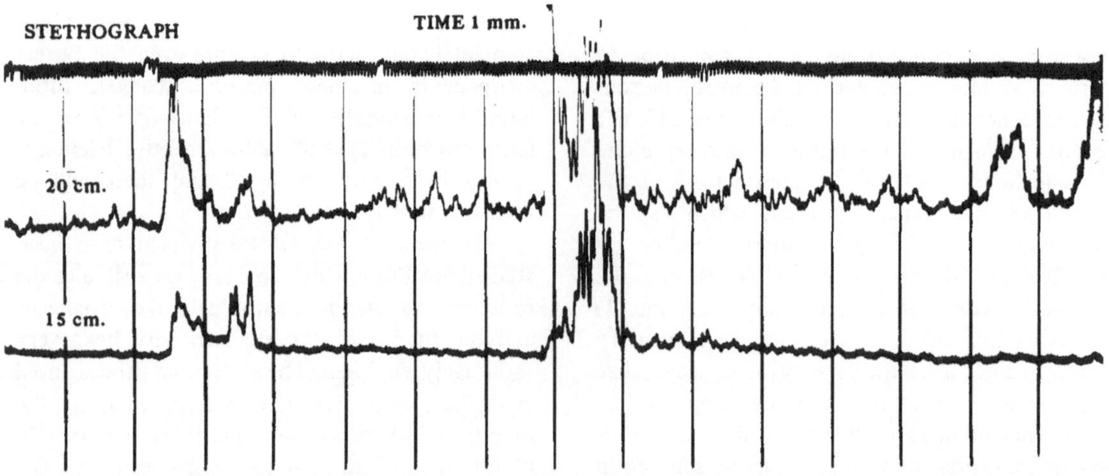


FIGURE 1. Example of normal colonic activity. Upper trace: stethograph. Middle trace: colon motility 20 cm from anus. Lowest trace: colon motility 15 cm from anus. Verticals = 1-min intervals.

information that one gets about motility; they increase comfort and allow the exploration of otherwise inaccessible areas of the alimentary tract. The increased information about motility devices derives from the fact that the capsule measures not only the pressure, but also the signal strength. This is a function of movement. Figure 2 is a recording from the stomach of a volunteer patient. The pressure record shows very little pressure change throughout this period of recording. But from the signal strength record it is clear that there are phasic changes recurring at approximately three per minute. In this situation a wave of contraction moves along the pyloric antrum, tilting the capsule as the wave passes under, allowing a change in the signal. The wave, however, is not sufficiently strong to alter the pressure in the antrum. The reverse situation occurs in the colon (Figure 3). Here we have a pressure record showing typical two-per-minute phasic changes in the sigmoid colon, representing pressures of around 40 to 50 cm of water. The signal strength record shows very little change at all. Thus, in the colon there is a great deal of squeezing and pushing with virtually no movement of the capsule.

Problems using these capsules have included their localization in the GI tract. One clue is to look at the nature of the record, but normally it is necessary to have some form of radiological monitoring of the position of the capsule. A small quantity of contrast medium given before the capsule outlines the intestine.

Temperature change in the alimentary tract considerably affects the base line. The temperature of the duodenum, which is close to the liver in the normal man, may be as high as 101°F. So base-line changes undoubtedly do occur along the alimentary tract.

Sometimes it seems we have instruments looking for some application. In digestive physiology and digestive diseases we have a large number of problems looking for a form of exploration. Large numbers of patients suffer from diseases of the alimentary tract related to some abnormality on the contractility and the contraction of the alimentary muscle, and frankly, we know very, very little about these disorders. These are very common complaints; they are ill-understood and ill-explored. One possible approach is to measure pressures in the alimentary tract. This is not the only approach and may not even be the best approach, but it is at least one which is feasible.

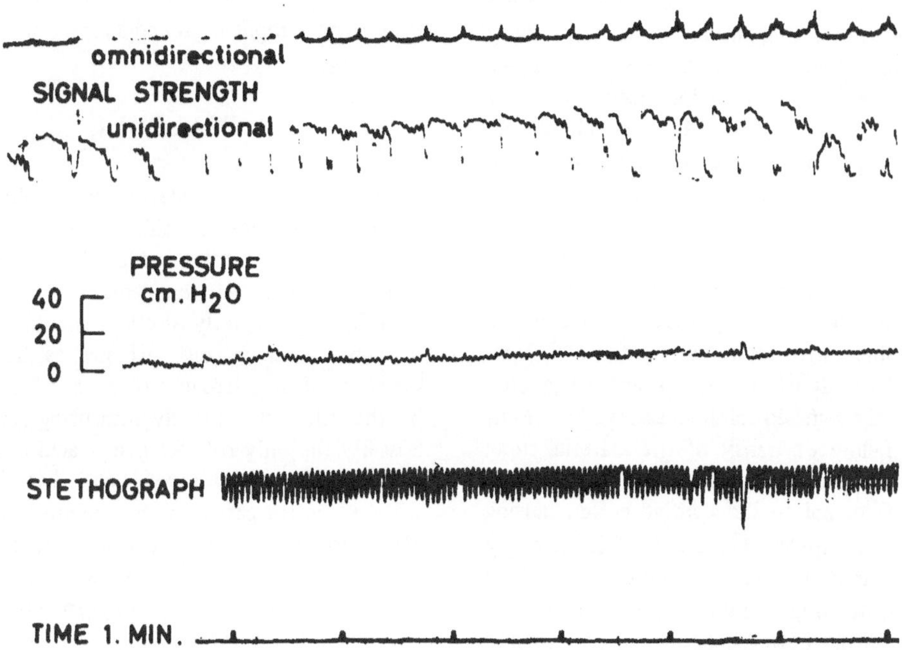


FIGURE 2. Unidirectional aerial shows regular three-per-minute movements of radio pill in stomach without any waves in pressure record. (From Connell, A. M., McCall, J., Misiewicz, J. J., and Rowlands, E. N., Observations on the clinical use of radio pills, *Br. Med. J.*, 2, 771, 1963. With permission.)

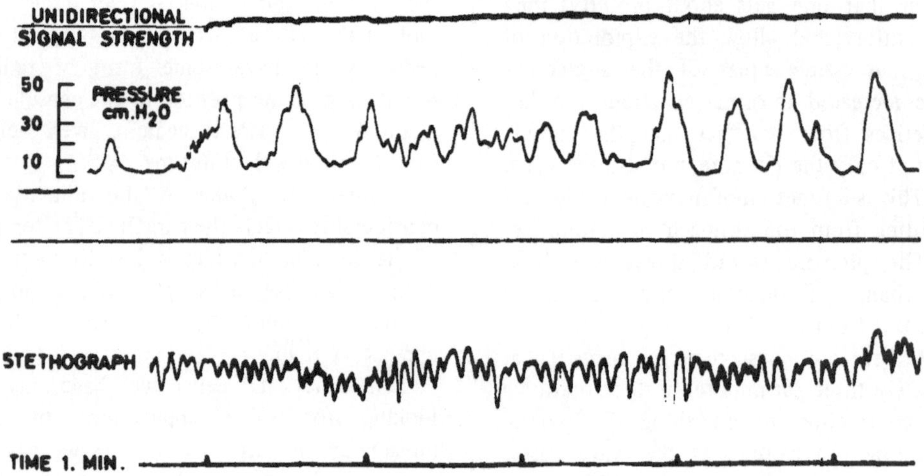


FIGURE 3. Pressure waves in colon without any movements of pill. (From Connell, A. M., McCall, J., Misiewicz, J. J., and Rowlands, E. N., Observations on the clinical use of radio pills, *Br. Med. J.*, 2, 771, 1963. With permission.)

It may be surprising to know that the largest cause of loss of time from work in this country is due to digestive diseases. Most digestive diseases involve the motor activity of the alimentary tract in some way. There are a large number of GI conditions which are extremely distressing and seriously affect the quality of life.

In one of our early studies with ingestible telemetering devices we observed a marked effect of the thought of eating on the segmenting activity of the terminal ileum. Eating increased the degree of activity. A similar situation occurs whereby eating affects the motor activity along the alimentary tract due to neurohumoral responses. Disturbances of these mechanisms produce a number of clinical situations. For example, many persons who have an irritable colon have symptoms immediately after eating. They feel distended or bloated. They may feel extremely uncomfortable, with cramping abdominal pain, and may have diarrhea. This is an exaggerated response of the sigmoid colon to eating. As a result there is a failure of transit of the material along the alimentary tract, and so there is distention elsewhere proximal to the sigmoid colon, causing the patient's symptoms (Figure 3). The only way we may obtain additional information about these common distressing conditions is through the development of better techniques of pressure recording in the dark areas between the duodenum and the distal colon.

There may be subtle, psycho, social, and political reasons why there is little thrust towards study of motor disorders of the GI tract. It appears as if the political thrust and much of the research thrust in this country and elsewhere in the western world are primarily concerned with the diseases that kill and the postponement of death. Perhaps we should concentrate more intensely on these conditions which are afflicting people in the prime of life. The type of afflictions discussed previously can be studied and understood, given application and adaptation of existing technology.

## DIALOGUE

**Dr. Geddes**: For instance, which physiological events would you consider most important to detect and transmit with endoradiosounds?

**Dr. Connell**: At this time we find pressure recording particularly useful. The pH recording may be of help, but pH probes have to be sensitive. Most gastroenterologists are disillusioned by the value of routinely measuring gastric acid. Possibly the only role for gastric acid is to provide a milieu for peptic activity, and there is perhaps still a place for gastric measurements. Only subtle pH measurments could give us some information of the metabolic processes in the small intestine. It may be useful to look at oxygen tension or gas tensions in the intestine.

**Dr. Geddes**: Why do you suppose the interest in endoradiosound disappeared so quickly when the technique was actively utilized in England and the United States, as well as in other countries?

**Dr. Connell**: I think it was a combination of a number of factors. The technical problems associated with using endoradiosounds are quite complex in clinical terms. There are logistic problems which have frustrated investigators and particularly clinicians. I think too that there is a problem in relation to the nature of the data in some areas. The radiosound[1,2] is moving on the crest of the pressure wave, so it does not give information about that particular wave. Probably it is only in the terminal ileum and in the colon, where the capsules are at rest for periods of time, that they are most useful. This brings me to a third point – this is an area of the alimentary tract that few clinicians are interested in, possibly because of lack of exposure to it.

**Dr. Fleming**: Your comments were fascinating! Could you tell us where the endoradiosounds get trapped?

**Dr. Connell**: Thanks for reminding me of that because it is quite important. I know of at least three patients who have had laparotomies following the use of radiosounds. In each case the radiosound unwittingly was held up behind a blockage which was not known to be there. Two of them, were carcinomas, one of them was Crohn's disease. These were undiagnosed lesions. So in a sense, the radiosounds performed a useful diagnostic service. I do not know of any cases in which the radiosound was held up in an alimentary tract that was otherwise normal.

**Dr. Fleming**: Does the radiosound ever get caught in the cecum?

**Dr. Connell**: I have not heard of it.

**Dr. Fleming**: Now for an additional question – Can you locate or possibly determine where the units are while in transit through the gut? If we were to use a radiosound, its location is important, and we might not be able to locate its relative position even when contrast medium is used.

**Dr. Connell**: There are two basic ways to determine the position of the unit. One is to put a small radioactive source into the radiosound and then perform a body scan; this procedure will tell you something about where the radiosound is located in space, although this is not really the information you would need to know. The other method is to simply have a shadow of contrast medium. We give 2 oz of barium 15 min before giving the radiosound, which helps to outline the cecum. The main difficulty is to distinguish if the pill is in the ileum or the cecum. Small intestine pressure movement patterns may help but are not conclusive.

**Dr. Fleming**: Permit me to restate the question. Could other criteria, perhaps parameters measured simultaneously, provide other clues? You have already discussed the measurement of the pH and $PO^2$, as well as the electrical activity, which is very difficult to obtain.

**Dr. Connell**: Electrical activity should be subjected to further investigation. Its a fundamental physiological mechanism, and if one could measure spike potentials, I think it would be very useful. Intraluminal sensors may not pick up spike potentials quantitatively, but it is possible to measure successfully the basal electrical rhythms (BER) or slow waves from the lumen. A sensor which could pick up BER together with the pressure would be useful.

## REFERENCES


1. **Mackay, R. S.**, Endoradiosondes, *Nature*, 7, 1239, 1957.
2. **Kimoto, S., Watanuki, T., Hori, M., Suma, K., Nagumo, J., Ouchi, A., Takahashi, T., Kumano, M., and Watanabi, H.**, Endoradiosonde for bleeding detection, *Med. Electron. Biol. Eng.*, 2, 85, 1963.

# ARTIFICIAL ORGANS

**Raymond J. Kiraly, Gordon Jacobs, and Yukihiko Nosé**

Implantable pressure transducer applications in artificial organ research are particularly relevant to the artificial heart. Both assist devices and total artificial hearts are under development and evaluation in experimental animals. Performance determinations of implanted pumps require accurate pressure measurements with minimum compromise due to the presence of the pressure-measuring devices. The requirements are becoming more stringent as chronic experimental animals are being monitored for several months to evaluate physiological changes over this period. The availability of better pressure transducers will be of significant benefit to this research.

The Department of Artificial Organs at the Cleveland Clinic Research Division is currently conducting research on artificial kidneys, lungs (oxygenators), liver, and heart. Biomaterials which are necessary for all of these devices are also being studied.

This chapter will concentrate on the artificial heart,[1,2] both total replacement as well as assist devices. With a pump in the body as an implanted artificial organ, the input and output pressures as well as flow are the significant points of interest. Thus the pressures associated with these pumps are critical measurements.

In the venous system, on the entrance to the right side of the heart, there are low pressures on the order of a few millimeters of mercury with reference to atmospheric pressure. At the output of the right ventricle the pressures are approximately 20 mmHg. The blood passes through the lung and back into the left atrium, where the pressure is again very close to atmospheric. The output of the left ventricle is the systemic pressure. It is about 100 to 150 mmHg.

One device which is currently used clinically is the intra-aortic balloon pump.[3] By means of a catheter through the femoral artery, an elongated balloon is put into the thoracic aorta quite close to the heart. As the balloon is inflated with gas, the pressure in the aorta rises slightly because of the displacement of the blood. As the balloon is collapsed by removing the gas through the catheter, the pressure is reduced because of the reduction in volume. Because of the greater pressures involved, the left ventricle does 85% of the work of the heart. So, if there is a cardiac disease or a problem after surgery, one means of assisting the heart would be to cycle this balloon, synchronized with the natural heart beat. Thus, when the natural heart starts to eject in systole, the balloon is deflated to lower the aortic pressure. When the natural heart has gone into diastole or is relaxed, the balloon is inflated and the aortic pressure is increased, which increases the perfusion to the coronary arteries, most of which takes place during the resting state of the heart.

There is a limitation on the effectiveness of the intra-aortic balloon pump, as it only changes aortic pressure slightly. A more effective way to take over some of the work of the left ventricle is to take the blood directly out of the left atrium, or the left ventricle, into a pump and pump it directly back into the aorta.[4] In this way the work of the left ventricle can be reduced markedly because it could be made to eject into the pump at near atmospheric pressure.

Another type of artificial heart is a total replacement device used in cases where the natural heart is so diseased that the entire organ has to be removed, valves and all, and a double ventricle pump is then implanted into the chest of the patient and the four connections are made to the vascular system. The measurements that must be made when doing this type of research will be discussed in this chapter.

Figure 1 shows how a pneumatically operated artificial ventricle[5] would work. A rigid case contains a flexible inner diaphragm or bladder. By putting air pressure inside, we can eject the blood out through the outlet valve, and when we reduce the pressure, or use a slight vacuum, the blood will flow in. This is typical of how nearly all research on artificial hearts has been done to date, using pneumatic systems. However, one could readily visualize a mechanical device moving this ventricle, using an electric motor or a high-pressure hydraulic line with a pressure-volume transducer.

Figure 2 shows a typical air drive system. Again very simplified, it has pressure regulators with a pressure source and a vacuum source. A simple solenoid valve connects the ventricle between either the pressure or vacuum. It can be controlled at a fixed rate or, if it is used as an assist device, synchronized with the natural heart. If the latter method is used, then some feedback would result, either from a pressure measurement or possibly the EKG signal. The air systems are used since they are quite simple, inexpensive, and easy to control.

Calves[6] are the experimental animals that have been used in most of the work to date, although some of the early studies started out with dogs, and other animals have been employed periodically. One reason that calves are now used is because the hemodynamics associated with a 200-pound calf are quite similar to an adult human. Also, the chest size is about the same, although the configurations of the chest are different. Calves are readily available; they are relatively inexpensive and are quite easy to work with. One problem with using a calf, however, is that it is an infant animal and is quite susceptible to the trauma of the surgical implantations and also quite likely, especially in this part of the country, to develop lung infections and pneumonia. For these reasons, the animals used in the studies must be carefully selected.

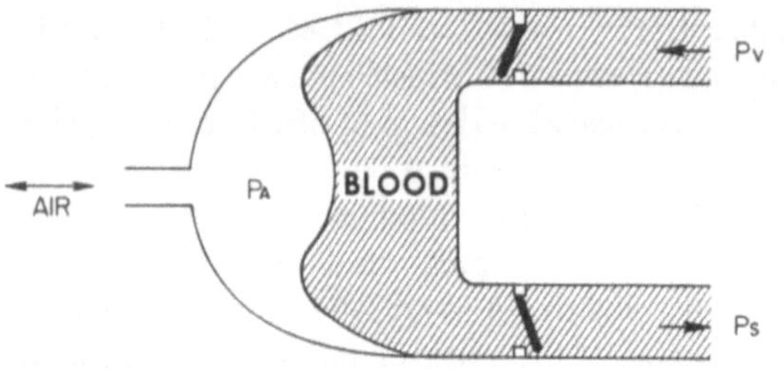


FIGURE 1. Typical representation of a pneumatically driven blood pump. The air and blood are separated by a flexible diaphragm. (From Kiraly, R. J., Jacobs, G. B., Urzua, J., and Nosé, Y., *Ann. Biomed. Eng.*, 4, 6, 1976. With permission.)

Let us next consider some of the actual artificial heart applications that are presently being planned or envisioned. Currently, hundreds of open heart surgeries are performed daily, mainly to by-pass occluded coronary arteries or for valve replacement. These procedures have quite a low mortality rate, but there are a significant number of patients who cannot be weaned off of the heart-lung machine in a very short period of time after surgery. So one of the applications would be to use an assist device for these patients temporarily, a matter of days or weeks until the natural heart has recovered sufficiently so that the device could be removed. The National Institute of Health has granted permission to the Texas Heart Institute and also to a group of hospitals in the Boston area to proceed with the temporary assist

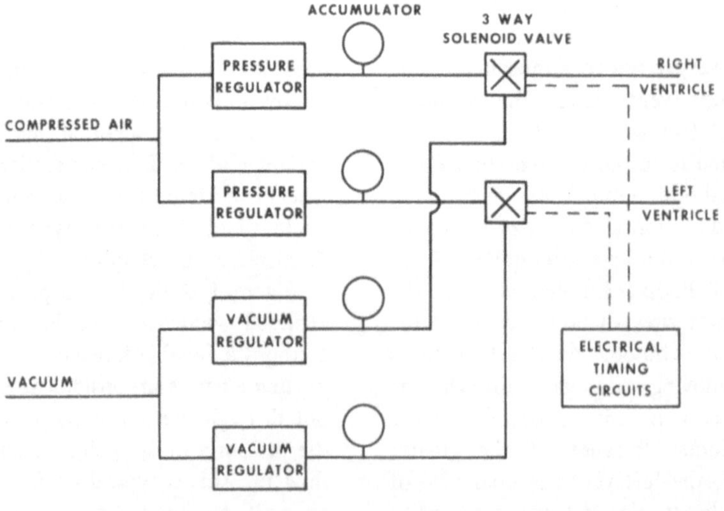


FIGURE 2. Schematic of a pneumatic driving system used for a total artificial heart. For single ventricle assist pumps, only half of the components are required.

devices when they have a patient that the doctors deem is a good candidate for using this device. The candidates would be those who have just undergone open heart surgery and cannot be taken off the external heart-lung machine. Long-term implantation is necessary if there is irreparable damage to the heart muscle. But, if the valves are in good shape, one might want to implant an assist device with a nuclear-powered system, which is totally implantable and useful for about 10 years. Total replacement, short-term support, could be an aid to transplantation. In 1969 a total artificial heart supported a patient for 3 days[7] while waiting for a donor, and this was somewhat successful in that case. After this pioneering effort, it seems that total replacement devices may serve both as a short-term support for transplantation and also as a permanent implant. The total artificial heart will be a very viable alternative to transplantation when it is determined how to make the artificial heart work for long periods of time.

Consider now how an assist device can be used with a natural heart. Connections can be made to the left atrium, where the blood is drawn into a pump and ejected into the aorta.[8] Another more convenient method that is now being used is to put a large bore cannula, approxiamtely 25 mm in diameter, right into the apex of the left ventricle and take the blood out into the by-pass pump.[9-11] The pumping then is synchronized with the natural heart. For example, using an air-powered system, with a slight vacuum in the pump simultaneous with systole of the natural heart, the heart would pump into this low-pressure source requiring little energy. Once the systolic duration of the natural heart has been completed, one can pressurize the air in this pump and pump it into the aorta, and thus reduce the work of the heart by a very large percentage.

For synchronization, one either measures the pressures or the EKG. Pressures in the pump, the aorta, and the left ventricle are measured. An electrical connector is used to pick up the EKG, and if there are major changes in the heart rate, then the system should be readjusted. In that case, having the pressure measurements would be beneficial. Of course, in a system like this there are connections to the left ventricle or the left atrium and to the aorta. The pump could be outside or inside the body. The same holds for the driving system and power source: in or out of the body. Possibly the pressure transducers can be incorporated right into the construction of the device.

Figure 3 shows the measurements we made in a 200-lb calf with the left ventricular assist device. The first record shows the air drive pressure. The peak pump flow is about 50 l/min, while the average flow is approximately 10 l/min. In this case the peak left ventricular pressure is obviously less than the minimum aortic pressure, so that it is known that one is by-passing the ventricular output and the natural aortic valve never opens. The aortic pressure spike observed is out-of-phase with the natural left ventricular pressure because the pump is ejecting while the natural heart is in diastole. When the pump is shut off, near normal aortic and ventricular pressures are again present. The effect of using the pump is that the area under the curve, which is the Tension-Time Index, has been reduced, thus markedly reducing the work of the left ventricle while increasing the average aortic pressure. For these types of experiments, it is very important to measure these pressures with implanted tranducers, one in the natural left ventricle and one in the aorta. Accurate and consistent measurements are very critical in these research and development efforts.

Figure 4 shows some of the problems encountered in the use of the total artificial heart. Plotted here, in a qualitative scale, is the relative mortality rate or the cause of death for animals depending on their survival time. Approximately

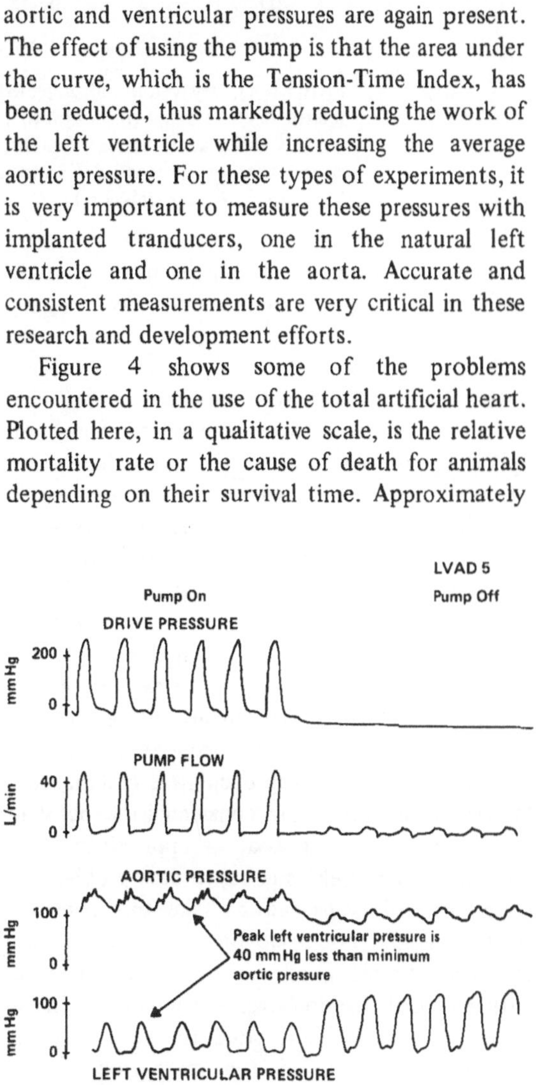


FIGURE 3. Hemodynamic parameters recorded for an assist pump in a 90-kg calf. The intrathoracic pump is connected between the natural left ventricle and the descending thoracic aorta. When the pump is "on," all of the cardiac ouput enters the pump.

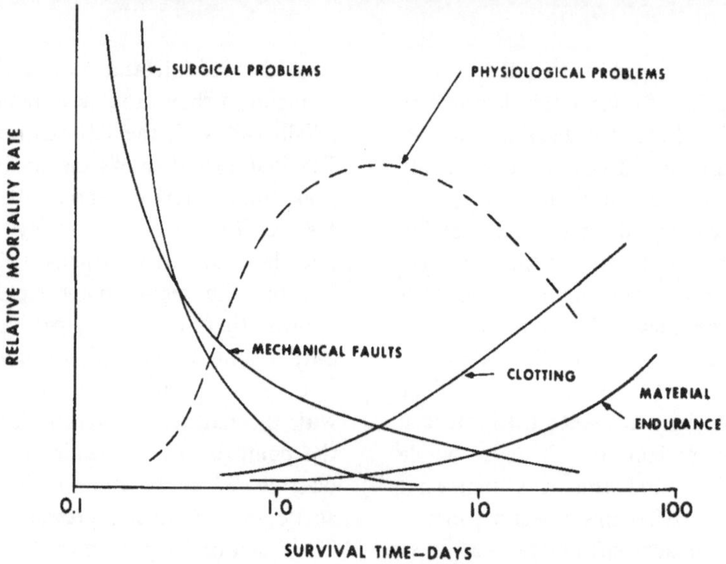


FIGURE 4. Historical and projected cause of death in experimental animals as related to the survival times with total artificial hearts. (From Nosé, Y. and Kolff, W. J., Cardiac assist devices, total artificial heart and unconventional dialysis, in *Perspectives in Biomedical Engineering*, Kenedi, R. M., Ed., Macmillan, New York, 1973. With permission.)

20 years ago when total artificial heart research had started, experimental animals were dying mainly because of the surgical and management problems associated with the new procedure. Actual survival times were measured in terms of hours. Also, the devices themselves were not operating properly; thus a series of mechanical or hardware problems existed. Approximately 5 years ago, survival times on the order of days were reached and new problem areas became apparent. These were physiological problems associated with the implanted pumps: problems of learning how to operate them effectively, how to match the proper anatomy, and how to control the devices. As we go on, we now see new problems on the horizon, and in fact, they are already being experienced as survival times are now approaching several months. Material endurance, if one considered the fact that the natural heart is beating approximately a million times a week, is going to be a problem. A good material is needed to take this repeated flexing. Also, blood clotting is becoming an increasing problem, as is probably experienced with endothelialization over the implanted transducer. It can be appreciated that this may be progressive with time and that more problems are coming up in this area. However, we already have some ideas on solving these last two problems, and we are still working mainly on the physiological problem areas with the total artificial heart.

One such physiological problem has to do with pressure measurements in the atrium. When the natural heart is replaced with an artificial device, one may compromise somewhat the natural anatomy, either by compressing the atrium or by adding a rigid structure.[12] The net result is an altered effectiveness of the atrium to serve as an accumulator for the low pressure blood. What happens with a small atrium, of course, is that it must distend as the blood fills it, with the net result being high osscillatory pressures measured in the atrium. With right atrial pressure fluctuations we observed damage to the liver, and with fluctuations in the left atrial pressures damage to the lungs was evident. In order to measure atrial pressures in this range, we have generally used fluid-filled catheters going inside of the body. Although we have not had much experience with implanted atrial transducers, we would like to use them if some of their problems can be solved.

We are currently studying the increase in the central venous pressure (CVP) of implanted animals. With short survival times, animals universally exhibited elevated CVPs; however, it now appears that after a few weeks, CVP may return to normal. The reason for these changes is presently being considered. Even a few millimeters change in the right atrial pressure is significant, and right now we are measuring these changes with catheters.

Control features of an artificial heart are related to the Frank-Starling relationship for the natural heart. This relationship simply states that the cardiac output is a strong function of the atrial pressure and that small changes in the atrial pressure will result in large changes in cardiac output. A normal heart might have a very steep slope, whereas a weak or failing heart has a shallow slope. What we try to do with the total artificial heart is mimic these characteristics. Here a steep slope is desirable, but it may turn out that it is not possible, especially after a long period of time where there are blood element deposits over the surfaces of the pump. With a total artificial heart, it may be necessary to use a pressure transducer to measure atrial pressure to control the pump-engine system.

Figure 5 shows a total artificial heart. One of the things that we are doing to lick the material endurance problem is working with the Goodyear Tire and Rubber Company developing rubber which has a high flex life. In addition, we are lining the inside of the device with natural tissue or protein coatings, chemically treated, and we are also using natural tissue valves both in and out of these devices. This is unique for artificial devices,[13,14] although tissue valves are routinely used in patients. One of the benefits of this system is that the animals can be maintained with no

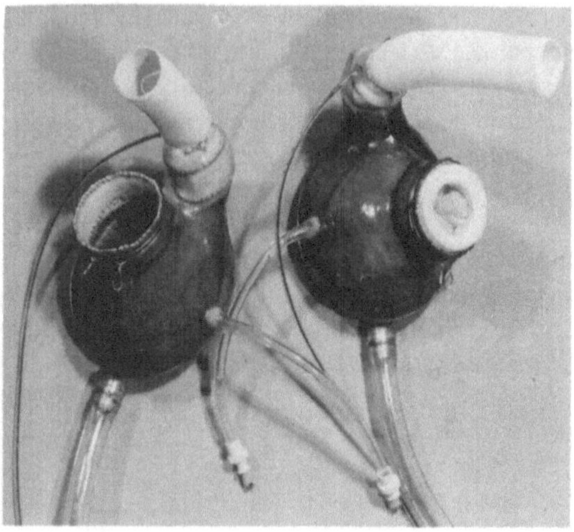

FIGURE 5. Photograph of both ventricles of a total artificial heart used in animal experiments. All blood-contacting surfaces are either chemically treated natural tissue or protein-coated. The right ventricle has the inflow valve in place.

anticoagulation after the surgical implantation. The fate of these surfaces is being studied in vivo.[15] We are also using the total heart to validate, in vivo, a mathematical model[16] which we are currently developing for an artificial ventricle combined with the natural cardiovascular system.

Figure 6 shows an animal having a total artificial heart. Here the two air lines coming out of the chest are visible. There are also other fluid-filled lines going to the external pressure transducers. Heparinized saline is continually pumped through these lines. Some of the problems encountered are related to maintaining the animal like this for a period of months. We have to keep monitoring the equipment, making sure the infusion pumps are working, and periodically flushing the catheters, as they still form clots especially on the outside of the catheters. The amount of heparin being added is small to affect the total animal, but it is preferable to have no anticoagulants in the animal. We also have an implanted flow meter. The pressures being measured are the right and left atrial pressures, the aortic pressure, and the pressures in the air side of the artificial heart. One problem with the tubes going through the skin is that they are, of course, a source of infection. Also, the tubes have to be somewhat rigid and not compliant. Therefore, as they go through the skin, they actually may force some anatomical changes in the device implanted in the chest of the animal or else interfere with the natural lung. Also, the level of the transducers must be adjusted for the hydrostatic reference point. The fidelity of the measurements is not as good as we would like it to be.

We let the chronic animals with assist pumps free in a large pen 5 by 10 ft in linear dimensions. In so doing we have only a single air line connected to the animal. We have investigated this with normal animals and found that if more than one line is used, kinking problems develop unless very elaborately designed swivels and brushes are employed to pick up these signals. So what we would like to do eventually is to run with an air line only and possibly use telemetry or periodically make connections to the implanted pressure transducers.

Figure 7 shows a calf with an assist device. The little transmitter on the harness is transmitting EKG to an FM radio that is approximately 5 ft away. The radio output is used to synchronize the

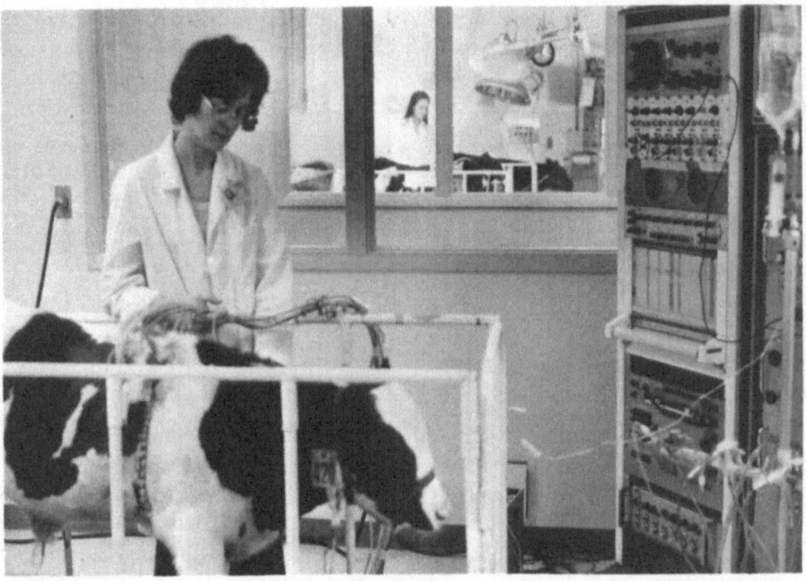

FIGURE 6. Photograph of animal with a total artificial heart. The fluid-filled catheters are connected to the pressure transducers mounted on the bar shown in the right foreground of the photo. Note the large number of tubes entering the animal's chest.

assist pump with the natural heart. We have also implanted pressure transducers and a flow meter that are periodically connected by wire to the recorders. However, it will be some time before we can really afford to use telemetry systems for pressure.

Another problem encountered with the implanted pressure transducers is that they are very susceptible to handling damage. We have built a special insertion tool so that the surgeons could put them into the left ventricle and aorta. Electric fields are very hazardous, and some transducers were lost because we turned on the electric knife at the time that the transducers were in the surgical room. We lost three transducers at that time, which is significant since they are quite expensive.

We have made special calibration equipment; the transducers are placed in a sealed box and pressurized. Zero shift is a problem and that is one reason that transducers are not being used for atrial pressure measurements.

Tissue overgrowth may be a problem. Recently, we sacrificed two calves. One calf had two transducers for 1 month and the other had two transducers for 2 months, implanted in the left ventricle and the aorta. The transducers in the calf for 1 month were both very clean. The 2-month animal had a clean transducer in the aorta, but the one in the ventricle had about 3 to 5 mm of tissue overgrowth. None of the four transducers were operational after the period of implantation.

We have tried to briefly indicate some of the applications and problem areas related to pressure measurement. One may look forward to advances in implantable transducers since accuracy, stability, and cost improvements will significantly beneift our research efforts.

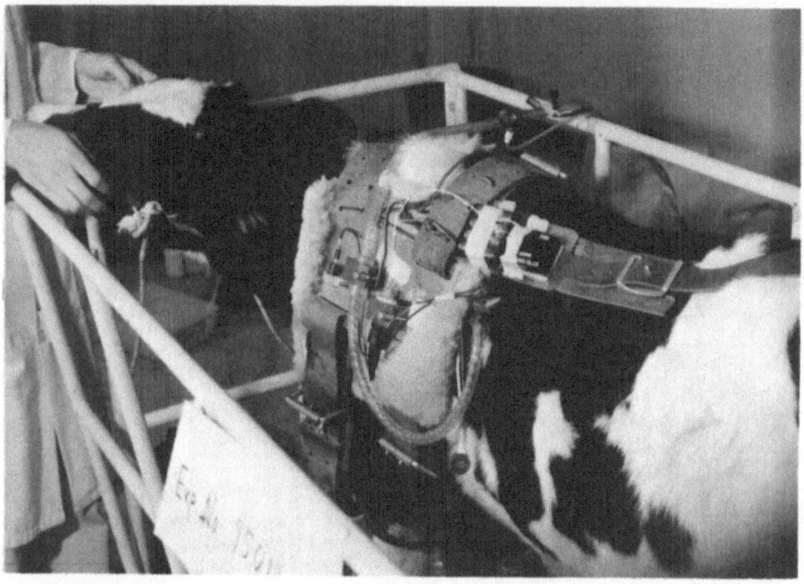

FIGURE 7. Photograph of animal implanted with left ventricular assist pump. The harness assembly absorbs stress from the air drive tube and instrumentation cables. EKG transmitter is also mounted on the harness. FM receiver in nearby control console allows synchronization of drive system with R wave.

## REFERENCES


1. **Nosé, Y. et al.**, Development and Evaluation of Cardiac Prostheses, annual report to Natl. Heart and Lung Inst., Contract NO1-HV-4-2960, May 1975.
2. **Nosé, Y. et al.**, Development and Evaluation of Cardiac Prostheses, annual report to Natl. Heart and Lung Inst., Contract NO1-HV-4-2960, March 1976.
3. **Weber, K. T. and Janicki, J. S.**, Intraaortic balloon counterpulsation, *Ann. Thorac. Surg.*, 17, 602, 1974.
4. National Heart and Lung Institute, Report on the Left Ventricular Assist Device, DHEW Publ. No. (NIH)75-626, January, 1974.
5. **Kiraly, R. J., Jacobs, G. B., Urzua, J., and Nosé, Y.**, Performance analysis of pneumatically driven blood pumps, *Ann. Biomed. Eng.*, 4, 6, 1976.
6. **Weber, K. T., Dennison, B. H., Fuqua, J. M., Speaker, D. M., and Hastings, F. W.**, Hemodynamic measurements in unanesthetized calves, *J. Surg. Res.*, 11, 383, 1971.
7. **Cooley, D. A.**, Transplantation versus prosthetic replacement of the heart, *Biomater. Med. Devices Artif. Organs*, 3, 481, 1975.
8. **DeBakey, M. E.**, Left ventricular bypass pump for cardiac assistance, *Am. J. Cardiol.* 27, 3, 1971.
9. **Bernhard, W. F., Poirier, V., LaFarge, C. G., and Carr, J. G.**, A new method for temporary left ventricular bypass, *J. Thorac. Cardiovasc. Surg.*, 70, 880, 1975.
10. **Robinson, W. J., Daly, B. D. T., Hughes, D. A., Migliore, J. J., Hibbs, C. W., Igo, S. R., and Norman, J. C.**, An abdominal left ventricular assist device, *Ann. Thorac. Surg.*, 19, 540, 1975.
11. **Nakiri, K., Morinaga, N., Washizu, T., Arancibia, C., Brown, D., Kiraly, R., and Nosé, Y.**, Left ventricle connection technique for cardiac assist devices, *Proceedings of the 28th ACEMB*, Vol. 17, The Alliance for Engineering in Medicine and Biology, Chevy Chase, Maryland, 1975, 350.
12. **Urzua, J.**, Physiology of the artificial heart, *Biomater. Med. Devices Artif. Organs*, 3, 161, 1975.
13. **Kiraly, R. J. and Nosé, Y.**, Natural tissue as a biomaterial, *Biomater. Med. Devices Artif. Organs*, 2, 207, 1974.
14. **Kambic, H., Picha, G., Kiraly, R., and Nosé, Y.**, Application of aldehyde treatments to cardiovascular devices, *Trans. Am. Soc. Artif. Intern. Organs*, 22, in press, 1976.
15. **Picha, G., Helmus, M., Barenberg, S., Gibbons, D., Martin, R., and Nosé, Y.**, The characterization of intima development in left ventricular assist device (LVAD) and total artificial heart (TAH), *Trans. Am. Soc. Artif. Intern. Organs*, 22, in press, 1976.
16. **Jacobs, G. B., Mitamura, Y., Sullivan, H., Kiraly, R., and Nosé, Y.**, In vivo characteristics of an artificial ventricle, *Med. Instrum.* (Baltimore), 10, 76, 1976.

# INTRACRANIAL PRESSURE MONITORING IN NEUROSURGERY*

**A. Earl Walker, Lawrence J. Viernstein, and John G. Chubbuck**

A pressure-sensing device containing a passive r-f resonant circuit, inductance, and capacitance whose natural frequency is influenced by the pressure of the environment has been developed for implantation in a burr hole in the human skull. The environmental (extradural) pressure is measured by observing the frequency at which the sensor absorbs energy from the magnetic field of a nearby grid-dip oscillator. The transensor has remained functional in monkeys and cats for periods of over a year. The base-line drift is approximately 15 mm of $H_2O$ per month. The pressures recorded vary with and at a level about 15% lower than the measured intraventricular pressures. Histological studies of the tissues adjacent to the implant have shown no cellular reaction.

## INTRODUCTION

Early determinations of the cerebrospinal fluid pressure (CSFP) were made by lumbar puncture or by introducing a catheter into the ventricle. Neither of these techniques is suitable for prolonged recording of the CSFP because of the danger of infection and patient discomfort. The value of continuous recording was first demonstrated by Lundberg,[7] who found that single CSFP determinations give inadequate and even misleading information due to the rather great dynamic variations in pressure. He described A, B, and C pressure waves in the CSF circulatory system. The A waves are now called plateau waves because of their characteristic shape. Hulme et al.,[6] in a study of 75 patients with space occupying lesions, showed that the plateau waves were pathological and represented a partial breakdown of pressure compensation in the CSF system. Hayden et al.[5] continuously monitored the intracranial pressure (ICP) in 25 hydrocephalic children and found a cyclic pattern of pressure with spontaneous episodic increases, especially at night.

The need for measurements of intracranial pressure resulted in the development of a number of different techniques. The most widely used method is that of an implanted capsule containing strain gauges. These devices require wire connections that pass through the scalp and, in some models, a snorkel tube for calibration. Their greatest source of error is a shift in the zero reading or base-line drift. These techniques have been quite satisfactory for short periods of time. However, the danger of infection, the risk of contamination by noxious substances if a break occurs in a capsule, and the base-line drift have deterred their long-term use.

Attempts to eliminate external connections entirely have been satisfactory only for short periods of recording. A number of ingenious techniques have been employed. The method to be described in this chapter for continuous monitoring of ICP requires only a passive transensor that electrically consists of an inductance and capacitance. The basic concept has been known for over 15 years. These transensors have been used for measuring intraenteric pressure in man,[3] bladder pressure,[4] intraocular pressure,[2] and intracranial pressure in monkeys.[8] However, because of their composition of Mylar® plastic and of glass, these transensors were found to be unstable.[1,9]

## SPECIFICATIONS OF THE MONITOR

The present intracranial pressure sensing implant (transensor) contains a passive r-f resonant circuit, inductance, and capacitance whose natural frequency is influenced by the pressure of the sensor's environment. Specifically, it compares the surrounding environmental pressure to that of the air entrapped inside the sensor. The environmental

*Supported in part by Grant NS 11710, National Institute of Neurological and Communicative Disorders and Stroke.

pressure is measured by observing the frequency at which the sensor absorbs energy from the magnetic field of a nearby grid-dip oscillator.

This principal of operation requires three important design features: (1) sufficient elastic compliance that a measurable deformation will result from small changes in intracranial pressure; (2) an extremely nonporous enclosure for the entrapped air so that it will not leak during the useful life of the implant; and (3) an electrically nonconductive implant so that the resonant circuit will not be shielded from the monitoring radio frequency magnetic field. Plastic enclosures that have been used in the past are ideally compliant and nonconductive, but much too porous. Glass enclosures are quite nonporous and nonconductive but, with walls of adequate thickness, very noncompliant. In the present transensor, the three requirements are met by sharing the enclosure surface area between ceramic, which is nonporous and nonconductive, and metal which is nonporous and compliant; because of the small area of the metal, its conductivity does not shield significantly the implant's field.

## TRANSENSOR

The component parts, except for the inductance coil, are shown spread out in Figure 1. Two ceramic "half-sections," when bonded together, form a hollow circular cylinder. The reference pressure enclosure is then completed by bonding a gold-plated nickel bellows into the central hole of each ceramic half-section as is shown in Figure 2. Both the bellows and ceramic sections are epoxy bonded. Epoxy is porous, however, the glue joints have a very high ratio of path length to diffusion cross-section compared to that of the ceramic portion of the enclosure and are, therefore, an insignificant consideration.

The reference pressure enclosure has an eight-turn inductance coil wound around a ceramic section. The coil ends are soldered to the bellows forming a passive resonant circuit in which the closed ends of the bellows (Figure 3) lie in close proximity to each other forming the capacitor portion of the circuit. An increase in pressure outside of the enclosure will elongate both bellows approximating the closed ends, thus increasing the electrical capacitance, and thereby lowering the circuit's r-f resonant frequency.

The conductive metal bellows shield approximately 20% of the magnetic field cross-section of the inductance coil. This shielding diminishes the coefficient of coupling between the implant and the external monitoring circuit, thereby reducing slightly the implantation depth at which the sensor's resonant frequency can be measured.

FIGURE 1. A breakdown of the transensor to show the component parts.

FIGURE 2. The transensor with the diaphragms in place.

FIGURE 3. The transensor assembled and ready for encasement.

The pressure sensing element is (Figure 4) enclosed in a plastic (polysulfone) case. The purpose of the plastic enclosure is to isolate the circuit from the conductive body fluid that would short-circuit the bellows capacitance. The two halves of the plastic case are cemented together using polysulfone dissolved in dichloroethane. The region outside the air enclosure and inside the plastic case is filled with medical-grade silicone liquid to transmit the pressure from the outer plastic case to the bellows. Silicone liquid was chosen for its very low dielectric constant, which

FIGURE 4. The encased transensor.

minimizes stray capacitance, for its very low moisture absorption (200 ppm), and for its lack of toxicity.

## CALIBRATION AND QUALITY CONTROL OF THE TRANSENSOR

Calibration of the implant is primarily a matter of determining the space between the closed ends of the bellows, and it is thereby inherently related to the reproducibility or quality control with which the sensors can be made. The design spacing of the bellows ends is 0.004 in. for standard atmospheric pressure and body temperature. The spacing uncertainty introduced by the bellows-to-ceramic bonds can be determined electrically and removed by honing the ceramic sections on a diamond dust-impregnated copper flat. Honing is done on the surfaces that are later bonded together. This procedure does not, however, remove the spacing error resulting from the uncertainty in thickness of the ceramic-to-ceramic bond. A technique of modifying the reference pressure to compensate for this spacing error has been successfully employed.

Control of the reference pressure requires the epoxy bonds to be formed and cured (baked) before final closure of the air entrapment, which is made by sealing a small vent hole at a specified temperature and pressure.

Compensation for error in bellows spacing requires an additional bias in the closure temperature at the rate of 10.8°C per milli-inch spacing error. With the presently used fabrication technique, the variation in bellows spacing due to the ceramic-to-ceramic bond is approximately ±0.4 mil.

The transensor is tested for gross leakage shortly after closing the vent hole by allowing the transensor to reach room temperature, during which time the resonant frequency diminishes. If gross leaks are present, the frequency will drift back to that present at body temperature. If the frequency remains stable at room temperature, indicating no gross leaks, the enclosure is then tested for a fine leak by being "soaked" in pressurized helium for several days so that the gas may seep through any microscopic leaks. The enclosure is then placed in the vacuum chamber of a helium tester to detect any helium that may escape.

## THE MODEL OF THE TRANSENSOR

The pressure ($P_o$) outside the ceramic case of the transensor is related to the pressure inside ($P_i$) by the following:

$$P_o = P_i + \frac{k}{A}(x_o - x)$$

where k is the spring constant of the bellows; A is the cross-sectional area of the bellows closed end; $X_o$ is the neutral (unstressed) position of the bellows (half the separation of the bellows); and X is the stressed position of the bellows.

Since the volume change caused by bellows movement is negligible, the internal pressure ($P_i$) can be expressed in terms of temperature as follows:

$$P_i = \frac{P_c}{T_c} T$$

where $P_c$ and $T_c$ are respectively the closure pressure and temperature at the time the vent hole in the transensor is sealed. The ideal values for these last two parameters are 10,336 mm $H_2O$ (760 mmHg) and 558°F (body temperature, absolute). Hence the temperature sensitivity of the transensor is 18.5 mm $H_2O$ per degrees Fahrenheit, and the formula giving the pressure correction ($P_T$) for body temperature ($T_B$) is

$$P_T = 18.5\,(T_B - 98.6)$$

The distance (X in milli-inch) of the bellows surface from the midpoint between the bellows can be given in terms of the resonant r-f frequency ($\omega$):

$$X = \frac{1/2\ \omega^2 LK}{1 - \omega^2 LC_s}$$

where L is the inductance ($1.4 \times 10^{-6}$ H); K is a constant (4.039); and $C_s$ is the stay capacitance ($\mu\mu$F). The general equation can now be written:

$$P_{BP} + P_A = \frac{P_c}{T_c} T + \gamma \frac{f_0^2 - f^2}{(1 - \beta f_0^2)(1 - \beta f^2)}$$

where $P_{BP}$ is the barometric pressure (mm $H_2O$); $P_A$ is the applied pressure (mm $H_2O$); T is the temperature (degrees Fahrenheit absolute); $\gamma$ is $(2\pi)^2 \frac{k}{A}\frac{K}{2}$ L; $\beta$ is $(2\pi)^2\ LC_S$; and $f_0$ is the frequency under standard conditions of pressure (10,336 mm $H_2O$) and temperature.

The last three numbers $\gamma$, $\beta$, and $f_0$ are sufficient to describe completely the transensor performance. The value for $f_0$ is measured directly and the values for $\gamma$ and $\beta$ are determined from two frequency readings, $f_1$ and $f_2$, at two different pressure readings, $P_1$ and $P_2$, made at a known temperature and barometric pressure. The theoretical and observed data are shown in Figure 5.

The monitoring system basically consists of a grid-dip oscillator that excites the resonant circuit of a pressure-sensing implant lying within its magnetic field. This provides a means of varying the frequency of the grid-dip oscillator so that it may be swept over a range of frequencies that includes the resonant frequency of the implant,

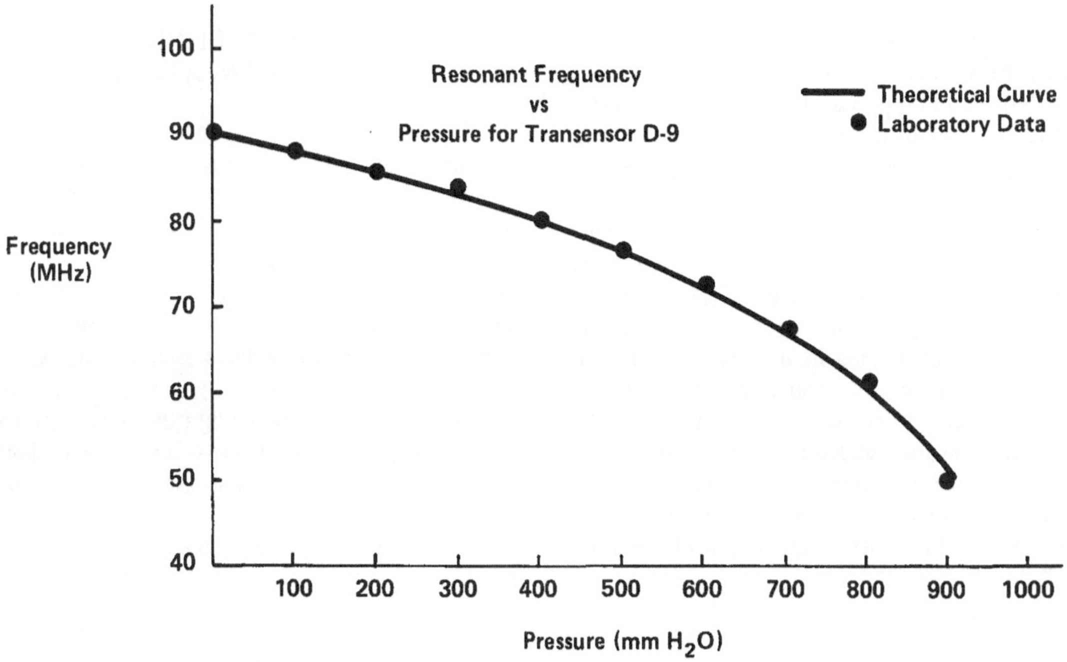


FIGURE 5. Theoretical vs. measured transensor response to pressure.

FIGURE 6. The detector unit (left), grid-dip oscillator (above), and curve with dip (below) shown on the cathode ray oscilloscope screen.

and some means of identifying the frequency at which the grid current dip occurs.

Briefly, a grid-dip oscillator is simply a resonant (high "Q") circuit consisting of a capacitor and an inductance coil which is driven by a low-gain active element such as a field-effect transistor. This combination forms an oscillatory system which has a very poor output regulation – its oscillation amplitude is markedly diminished by even the slightest loading, such as by the pressure transensor implant lying in the magnetic field of the oscillator inductance coil when the frequency of the oscillator coincides with that of the implant. When the oscillator is swept in frequency over the resonant frequency of the transensor, the loading is observed as a drop in the gate current drawn by the FET. Hence, the notion of a grid-dip oscillator (Figure 6).

## SURGICAL TECHNIQUE FOR IMPLANTATION

The sensor is implanted under local anesthesia through a curvilinear incision in the scalp. A standard air-driven trephine or brace trephine may be used to make the hole. Bleeding is controlled by bone wax and the electrocautery. The sensor is held securely in place by means of a collar affixed to the skull (Figure 7). The flange of the collar extends over the outer table and, through small holes, is fixed to the skull by screws or sutures. The sensor is screwed down into the collar until its inner face – the diaphragm – apposes but does not indent the dura mater so that the underlying subarachnoid space is not compromised. The desired depth can be monitored by a change in resonance frequency when the diaphragm makes

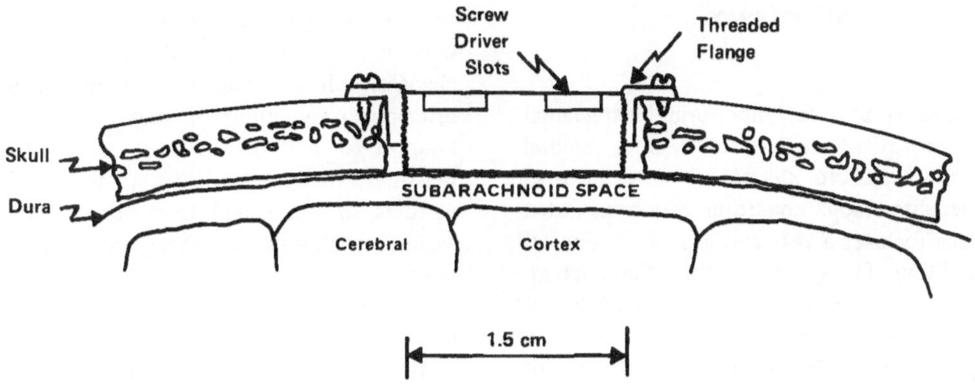


FIGURE 7. Sketch to show the technique for implantation and fixation of the transensor in the skull.

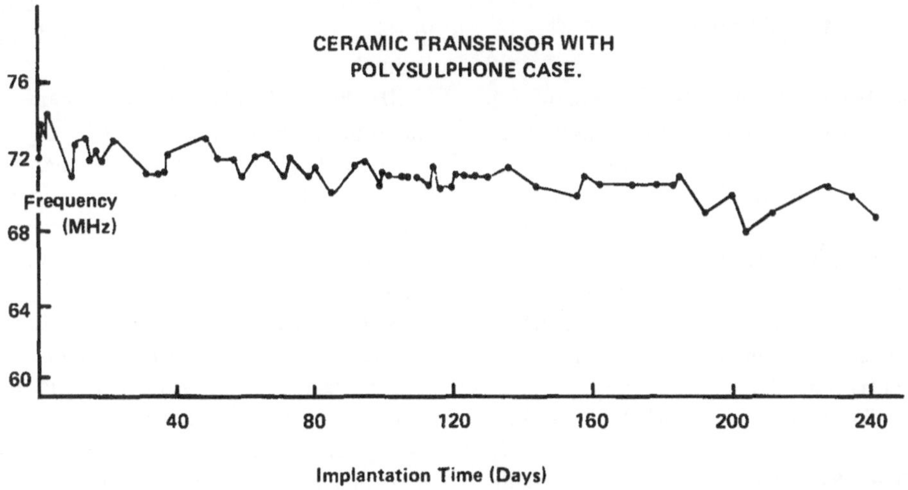


FIGURE 8. Graph of the resonant frequencies determined over a period of 240 days.

contact with the dura mater. The galea of the scalp is then sutured, and usually steritape is used to approximate the edges of the wound. A dressing may be applied or the wound may be sealed with a plastic spray. Dressings do not interfere with the operation of the interrogator unit which may be taped over the side of the sensor.

## PERFORMANCE

This transensor in monkeys and cats has remained functional for periods of over a year. However, the gradual diffusion of water through the plastic diaphragm of the transensor originally used caused a base-line drift of 75 to 100 mm $H_2O$ per month. These drift rates are acceptable for short-term monitoring. For long-term recording, a polysulphone plastic or parylene- coated Lexan® is used, which reduces diffusion to a level of 15 mm $H_2O$ per month. This base-line drift may be compared to the 450 mm $H_2O$ per month shift given by strain-gauge transducers. An example of a long-term recording is shown in Figure 8, which was taken from a transensor implanted in a cat.

In acute experiments in cats, the pressure as registered by the transensor in a burr hole was compared with the intraventricular pressure, which was raised manometrically to levels as high as 500 mm $H_2O$. Tracings made at various pressures demonstrated that the transensor closely followed variations in the intraventricular pressure. It was noted, however, that the pressure as determined by the transensor at the epidural location was generally 15% lower than the intraventricular pressure.

Gross and histological studies of the tissues at the site of implantation in monkeys have shown no reaction or other pathological change.

## SUMMARY

A simple system for monitoring intracranial pressure in patients with head trauma, cerebral edema, hydrocephalus, subarachnoid hemorrhage, and other intracranial conditions has been tested in animals for over a year and in man for shorter periods of time. The system consists of an implantable pressure transensor, an extracorporeal detector, and a bedside recorder. The device, having no wires passing from the scalp, has little risk of infection, and yet there is stability or little base-line drift over periods of months.


## DIALOGUE

**Dr. Fryer**: What is the diameter of the unit?

**Dr. Walker**: The diameter of the transensor is about the size of a nickel (12.7 × 6.4 mm).

**Question**: You showed a graph of 240 days recording, but you did not show the conversion in terms of millimeters of water.

**Dr. Walker**: The initial pressure would be approximately 100 mm, and the final pressure might be a little lower; there was a base-line drift of 0.408 mm $H_2O$ per day.

**Question**: Would you tell us what 68 roughly means?

**Dr. Walker**: Yes, that is a calibration which must be made for each transensor; in this case it represents approximately 2 mm $H_2O$ with reference to the original calibration.

**Mr. Konigsberg**: How many set units have been tested over that period of time exhibiting comparable stability?

**Dr. Walker**: We have, in monkeys and cats, approximately ten such implants; some of them had leaks and base-line drift, which came down within about 40 to 60 days. The rest of them were of the above type. We test all the implants for a period of 1 month before they are put into a patient or an animal. If, in that period of a month, there is no appreciable decrease in the frequency, we consider that the device will be stable. Usually a drift will show up in a period of a month. We have four patients in which the device has been implanted for periods up to 2 months.

**Dr. Fryer**: You mentioned frequency response and its importance. What is the duration of these cycles – minutes, hours, or days?

**Dr. Walker**: The cyclic changes to which I refer to in hydrocephalics, for example, occur perhaps once in a 24-hr period, usually at night. The other changes, which occur in patients with acute intracranial hypertension, may take place every hour or so, and they usually occur with higher pressures; the pressure goes up for a period of minutes or hours and then comes back down again. So-called plateau waves occur every 30 min or so.

**Dr. Fryer**: So the time scale is in the order of minutes?

**Dr. Walker**: In the present clinical use, we are measuring phenomena in the order of minutes or hours, but we can observe changes of seconds duration. We have short-term observations in patients in which readings were made every hour around the clock for a week.

**Dr. Fryer**: I understand that there were some variations in the order of seconds.

**Dr. Walker**: Yes, this is quite true. We have not looked at these short-term records in any detail. We are more interested in long follow-ups. We have to use a pen recording technique for short-term studies.

**Eddie Cheng**: What is the tissue reaction to the plastic? Can you do an in vivo calibration?

**Dr. Walker**: It is possible to calibrate in vivo, but we do not ordinarily do an in vivo calibration except in animals. Our assumption is that if the base line is stable over a period of months, its calibration will also be the same at the end of the experiment. This has been verified at the end of the experiment after the transensor has been retrieved. The first question relates to the reaction to the implantation of a substance such as plastic. We have looked for this reaction in our monkeys and in our human cases, but have not encountered it. I have no doubt, however, that there may be cases in which a rather severe connected tissue reaction may occur to an implant.

**Dr. Sandler**: Dr. Walker, I am surprised that you did not show some records that would demonstrate the comparison of the transducer with the external manometer or a CSF pressure transducer.

**Dr. Walker**: We have a comparison with transensor and intraventricular pressures in a cat. The points follow very closely with a difference of 10 or 20 mm of spinal fluid, the ventricular pressure being higher.

**Mr. Konigsberg**: Has there been any evidence of osmosis or pressure build-up within or decrease within the space between the Lexan® and the biological fluid? This is of great interest in other areas. Is that Lexan diaphragm impervious to migration of any molecules, whatsoever?

**Dr. Walker**: With some of the plastics there has been a build-up. With the plastic that we are now using – Lexan – there does not seem to be sufficient transmigration to constitute a problem.

**Mr. Konigsberg**: That seems to be a good substance?

**Dr. Walker**: Yes, that was one of the difficulties we had earlier when we were using Mylar®.

## REFERENCES


1. **Atkinson, J. R., Foltz, E. L., and Shurtleff, D. B.**, Radio telemetry for the measurement of intracranial pressure, *J. Neurosurg.*, 27, 428, 1967.
2. **Collins, C. C.**, A miniature passive pressure transensor for implanting in the eye, *IEEE Trans. Bio-Med. Electron.*, BME-14, 74, 1967.
3. **Farrar, J. T., Berkley, C., and Zworykin, V. K.**, Telemetering of intraenteric pressure in man by an externally energized wireless capsule, *Science*, 131, 1814, 1960.
4. **Gleason, D. M., Burbaum, C., and Lattimer, J. K.**, Pressure telemeter: A method of measuring pressure without tubes, wires, or catheters, *Surg. Forum*, 15, 502, 1964.
5. **Hayden, P. W., Shurtleff, D. B., and Foltz, E. L.**, Ventricular fluid pressure recordings in hydrocephalic patients, *Arch. Neurol.* (Chicago), 23, 147, 1970.
6. **Hulme, A., Chawla, J. C., and Cooper, R.**, Monitoring of intracranial pressure in neurosurgical patients, *J. Neurol. Neurosurg. Psychiatry*, 34, 108, 1971.
7. **Lundberg, N.**, Continuous recording and control of ventricular fluid pressure in neurosurgical practice, *Acta Psychiatr. Neurol. Scand.*, Suppl. 194, 1, 1960.
8. **Olsen, E. R., Collins, C. C., Loughborough, W. F., Richards, V., Adams, J. E., and Pinto, D. W.**, Intracranial pressure measurement with miniature passive implanted pressure transensor, *Am. J. Surg.*, 113, 727, 1967.
9. **Olsen, E. R., Collins, C. C., Altenhofen, T. R., Adams, J. E., and Richards, V.**, Intracranial tissue studies relating to glass transensors, *Am. J. Surg.*, 116, 3, 1968.

# SYSTEMS FOR THE LONG-TERM MONITORING OF INTRAVENTRICULAR PRESSURE IN NEUROSURGERY

**Ronald J. Lorig, Eddie M. Cheng, and Wen. H. Ko**

Two implantable intraventricular pressure transmitters have been built and tested for periods up to 8 months. The systems are designed to be incorporated with standard neurosurgical hardware for the treatment of intracranial pressure regulation problems. Features of the two systems include their stability in long-term testing, their very low sensitivity to body fluids, and the provision of a second transmitted data channel of *in situ* transducer temperature for thermal compensation. Stability has been good enough to permit long-term application in dogs for drug evaluation and physiologic investigation.

## INTRODUCTION

Intracranial pressure monitoring was established as a clinical tool by Lundberg (Sweden) in 1960.[1] Since that time, monitoring by indwelling intracranial catheter has been used in many neurosurgical centers for the management of postoperative and head trauma patients where intracranial space-occupying lesions may develop, causing nervous system dysfunction by compromise of local or global cerebral blood flow.[2] While this monitoring via indwelling intracranial catheter is valuable, it is limited to short-term (up to 5 to 7 days) use by the potential risk of infection along the catheter. Responding to the need for an improved long-term monitoring technique, the Engineering Design Center and the Division of Neurosurgery at Case Western Reserve have developed two completely different implantable capsules, each capable of transmitting intracranial pressure and temperature over periods of 6 months and perhaps longer.[3]

In contrast to techniques which measure epidural pressure over the short term, as described in the preceding section by Walker et al., the systems are designed to be incorporated with valve-regulated ventriculo-atrial or ventriculo-peritoneal shunts in patients, both infants and adults, who have hydrocephalus or other space-occupying lesions which must be palliated by pressure regulation. In our system, the subcutaneously positioned capsule is connected in parallel with and proximal to the valve, via a standard silicone rubber catheter. A catheter in communication with the ventricle (and the cerebrospinal fluid -- CSF) conducts pressure information to these devices. These elements are shown in the diagram of Figure 1.

We will now describe each of the two capsules which implement the measurement and transmission of pressure and temperature. These have been developed chronologically as the Pi system and the Si system. The Pi system uses the Pitran as the transducing element, while the Si system employs the EDC silicon transducer described in the Hynecek section.

## THE PI SYSTEM

The intracranial pressure (ICP) is normally less than 10 mmHg and may range higher than 50 to 100 mmHg. A commercially available pressure-sensitive transistor (Pitran -- Stow Labs, 2 psid) for this range is mounted in the kovar compartment of a cylindrical (10 mm O.D.) kovar-Pyrex® tube. A kovar cap and port are micro-laser welded to the cylinder to hermetically seal the transducer end; cerebrospinal fluid (CSF) accesses the pressure-sensitive diaphragm through the port, which can accept standard silicone rubber catheters.

The electronics which comprise the rest of the transmitter package are inserted through the Pyrex end of the capsule, which has been previously sealed with a Pyrex cap. The transmitter is a modular device and consists of subsystems illustrated in the block diagram in Figure 2. Power is derived from two sources in parallel. A battery-magnetic reed switch combination permits intermittent operation or continuous short-term operation. The battery is a 1.3-V AgO cell which has to supply 50 $\mu$A to the hybrid electronic

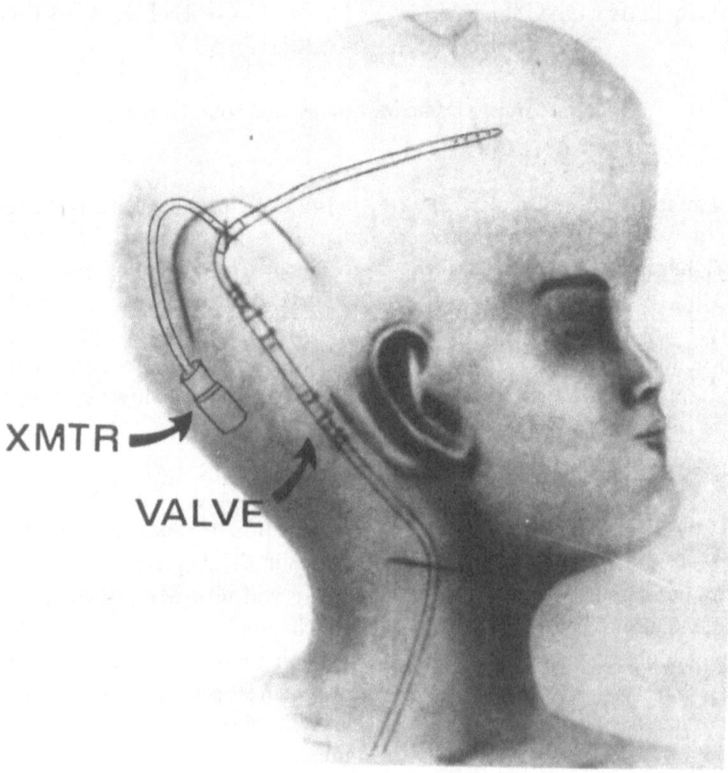


FIGURE 1. Elements of the implantable intraventricular pressure measurement system in association with standard neurosurgical hardware.

package. Also available is power from a radio-frequency inductive power source operating nominally at 3.5 MHz. Since coupling may be variable, a hybrid voltage regulator buffers the output of the LC resonant detector. Thus, the RF is available anytime and certainly in the event of battery failure, especially for very long-term usage. The remaining modules are signal processing for the pressure and temperature compensation systems.

The pressure transducer current is modulated by the applied pressure; the transmitter carrier is gated on and off by a pulse train whose period is proportional to the pressure. The carrier frequency (nominally ~100 MHz) thus consists of bursts of RF energy related to pressure. The subcarrier repetition rate is on the order of 1 to 10 kHz. In the same package, a temperature transmitter, also developed at the Engineering Design Center of CWRU, which operates at a slightly different carrier frequency has been included. While efforts have been applied to give some internal compensation for thermal effects, this added feature results in the greatest accuracy, since the *in situ* transducer temperature is readily known. The thermal and atmospheric pressure effects can thus be automatically corrected in the external receiving devices. Thus, the electronics described and shown in Figure 2 have been placed in the package of Figure 3. The entire package is coated by a multilayer, dipped application of segmented polyurethane. The coupling with the silicone rubber tubing at the transducer end is merely secured by a pair of silk ties. Sterilization by ethylene oxide has been found adequate in the animal tests made to date. There has been minimal evidence of any corrosion effects anywhere on the package; in tests over a 7-month implantation period, there has been no leakage to the electronics evidenced and base lines have remained stable enough to conduct long-term physiological investigation.

## THE SI SYSTEM

This system utilizes an in-house designed and

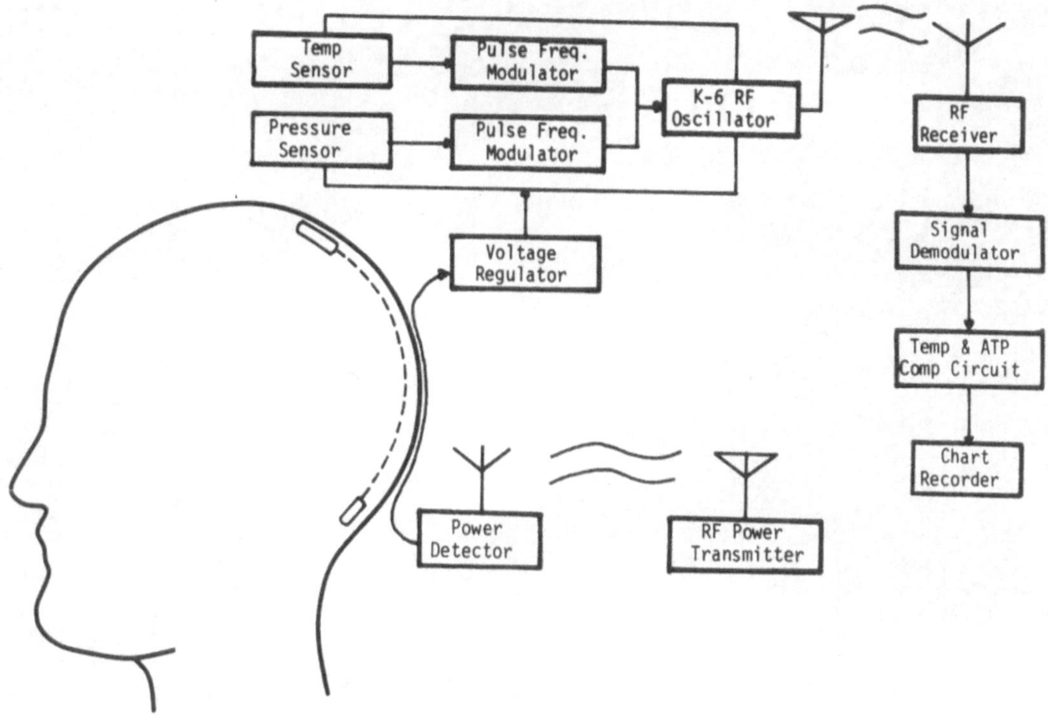


FIGURE 2. Block diagram of the modules of the implant, power, and receiver systems.

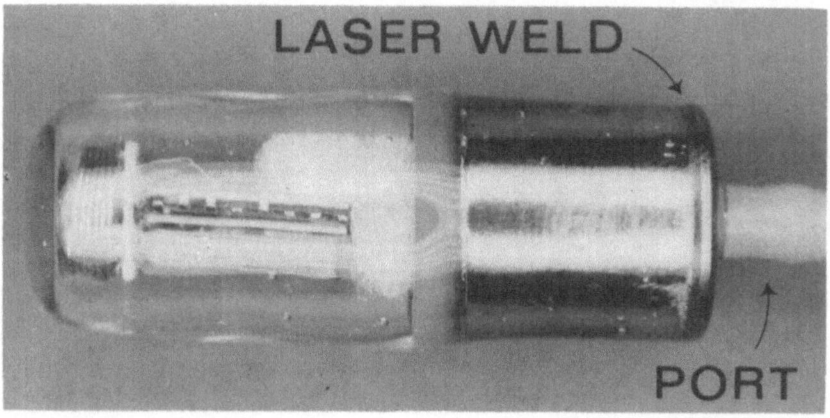


FIGURE 3. Photograph of the hermetically sealed glass-kovar package using the Pitran® transducer. Overall length is 4 cm.

fabricated silicon substrate pressure and temperature transducer, described elsewhere in the volume (see chapter by J. Hynecek). The transmitter sends out amplitude-modulated 120-MHz pulses in the form of low duty cycle groups. These pulse groups are frequency modulated by the temperature information. Within each group, the pulse position is modulated by the pressure information. A chopper technique is employed to eliminate the effect of thermal drift on the electronic components. All of the electronic components are put on a thick film circuit substrate which fits inside a 1 × 1/2 × 1/8 in. ceramic flat-pack which is subsequently sealed. Since all of the external connections to the terminals of the flat-pack are low-impedance, encroachment of body fluids to the terminals (including the transducer connections) does not significantly affect per-

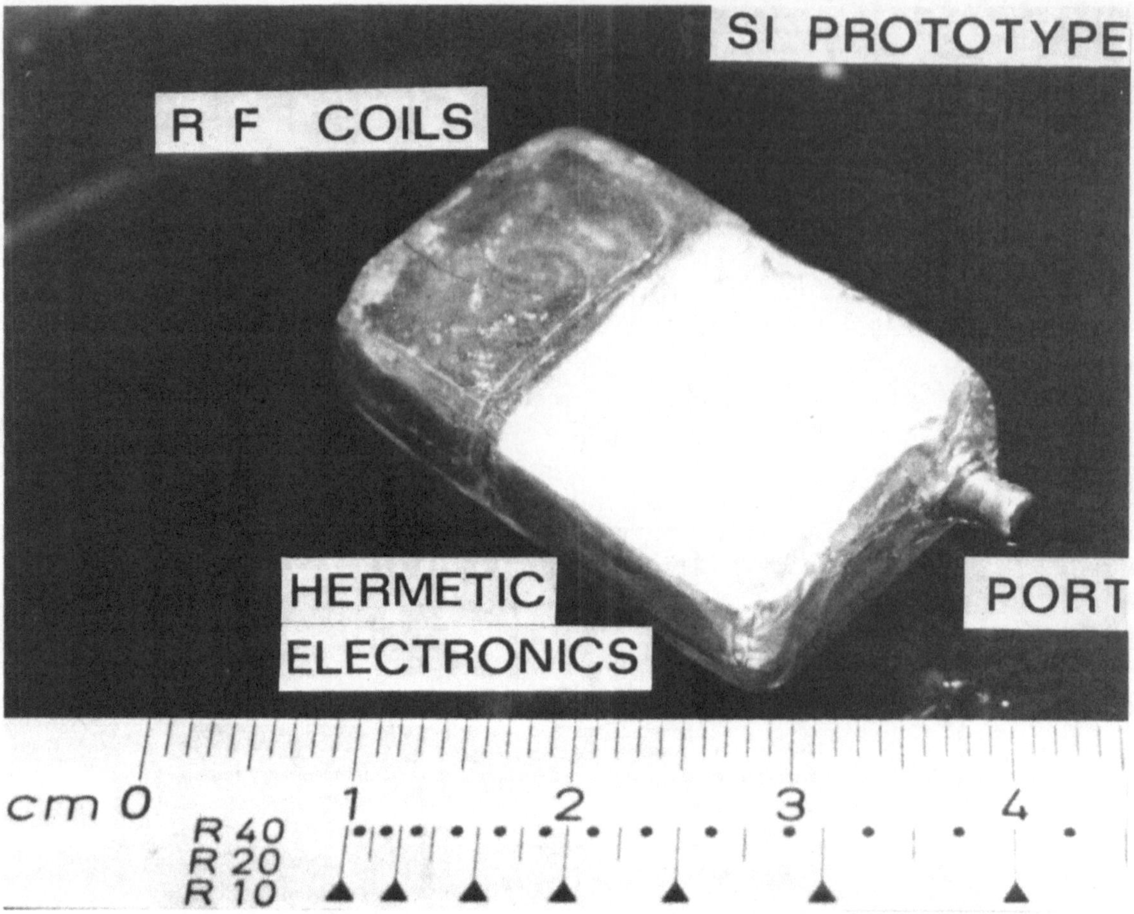


FIGURE 4. Photograph of the transmitter system using the EDC silicon strain-gauge pressure and temperature transducer. Overall length is 3.5 cm.

formance. At a temperature range of 37°C ± 3°C, the pressure signal accuracy achieved is ±1 mmHg. The pressure range of the implant is 300 mmHg.

This unit, merely coated by segmented polyurethane, has been implanted in two kaolinized* dogs. The stability has been maintained within 2 mmHg of the original base line for the greater than 2-month duration of the continuing implant tests. Figure 4 illustrates this recently developed advancement.

Overall then, inductive power at 3.5 MHz is beamed into either package, and pressure- and temperature-modulated RF is returned at 120 MHz. The receiving systems are good receivers operating in series with appropriate phase-locked-loop detectors and demodulators, the output of the system being both analog and digital readout of the intracranial pressure and transducer temperature.

Figure 5 illustrates placement of the package in a hydrocephalic dog. Visible is a standard Rickham® reservoir through which connection is made to CSF. The silicone rubber cap of the reservoir can easily and routinely be tapped by a needle to recalibrate the system if this is necessary. The package is loosely fastened to subcutaneous muscle, where it develops an entirely normal foreign body tissue reaction. In a patient, this apparatus would be accompanied by a parallel line such as a ventriculo-atrial shunt for pressure regulation. In that case, pressure under and over regulation would be detected, and there would be appropriate attention drawn before unnecessary clinical signs would become manifest.

*Kaolinzation of the *cisterna magna* causes an enlargement of the lateral ventricles of the brain and provides easier access to the CSF. The resulting condition mimics hydrocephalus.

FIGURE 5. X-Ray photograph of the placement of Pitran® transmitter system in a dog.

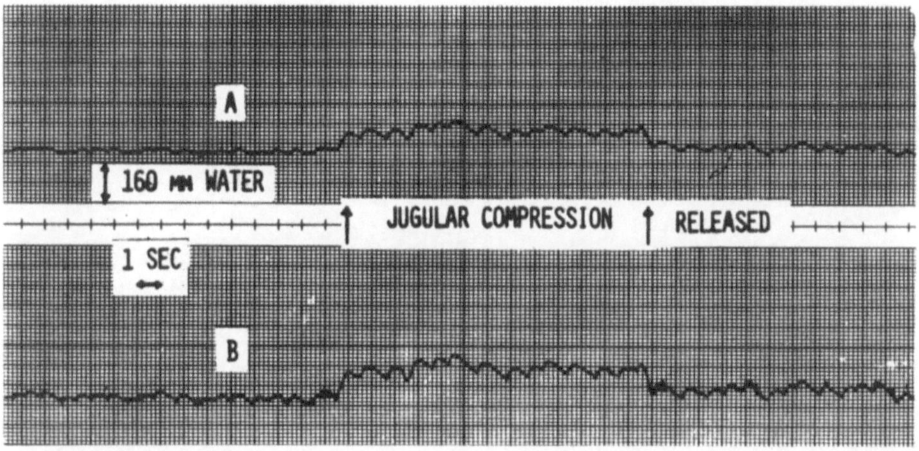


FIGURE 6. Comparison of transmitted intraventricular pressure (Trace A) by the Pitran® transducer system and direct (Statham® transducer Trace B) recording from the same ventricle through a Rickham® reservoir.

Since the pressure is sampled at a high rate, a continuous waveform can be obtained, and in fact, is band-limited to 30 Hz by external receiving equipment. While the mean level of ICP is usually the referred-to parameter, the capability to observe cardiac and respiratory activity serves some important purposes. For example, if these pulsatile activities are noted to attenuate or even cease, a possible cause might be advancing occlusion at the catheter tip; accompanied by a rise in mean ICP, increased pulsation amplitude may be an indication of failing compliance in brain to absorb expanding mass lesion.

Figure 6 shows a typical comparison of direct transducer measurement of ICP and simultaneous telemetry measurement during jugular compression (an induced means of temporarily increasing ICP by decreasing intracranial venous outflow). It is apparent that reproduction (to within a scale factor) is excellent, and in fact, avoids problems like motion artifact of the line leading to the extracorporeal transducer. The base-line stability of the system is good, about 2 cm $H_2O$/week; however, too few units manufactured in a similar manner have been evaluated, to date, to give a meaningful figure.

The implications of the availability of an implantable pressure telemetry system which is basically stable, and which can be recalibrated in vivo should the need arise, are manifestly numerous. Clinically, it is possible to monitor acutely ill patients beyond the usual 5 to 7 days. The need is even more apparent in patients requiring long-term observation for chronic conditions. The effects of drug therapy on intracranial pressure can be evaluated, and in the laboratory, long-term undisturbed physiological investigations become feasible and reasonable.

## REFERENCES


1. **Lundberg, N.,** Continuous recording and control of ventricular fluid pressure in neurosurgical practice, *Acta Psychiatr. Scand.*, 36, Suppl. 149, 7, 1960.
2. **Brock, M. and Dietz, H.,** *Intracranial Pressure: Experimental and Clinical Aspects,* Springer-Verlag, New York, 1972.
3. **Cheng, E., Ko, W., Lorig, R., Beveridge, J., Brodkey, J., and Nulsen, F.,** Intracranial Pressure and Temperature Telemetry System Using Asynchronous PFM's with RF Shifting for Multiplexing, Proc. 28th ACEMB, New Orleans, September 20–24, 1975.

# PRESSURE MEASUREMENTS IN OBSTETRICS

**Michael R. Neuman**

The measurement of pressures in various portions of the reproductive tract is important in clinical monitoring and in basic research in reproductive biology. Direct and indirect methods are used to monitor intrauterine pressure during the labor as a means of determining the frequency, relative strength, and duration of uterine contractions. The indirect measurement is based on tonometry through the anterior abdominal wall using a tokodynamometer. Direct measurements of intrauterine pressure can be made either by means of a fluid-filled catheter coupling the amniotic cavity to an external pressure transducer, or through the use of a miniature semiconductor strain gauge pressure transducer placed in the amniotic cavity itself. It is necessary to design this transducer and its package so that it is only sensitive to hydrostatic pressure and not to direct forces. There are errors associated with all of these pressure-measuring techniques, which result in amplitude and waveform distortion as demonstrated by the simultaneous use of three pressure sensors. It is better to average the amplitude readings from several transducers for an individual contraction than to rely on a single recording. Pressure measurements are also made in the nonpregnant uterus and oviduct using similar direct techniques; however the errors are often more serious.

Pressure measurements are important both clinically and in basic research in obstetrics and gynecology, as well as in reproductive biology in general. Applications of pressure measurements may be categorized in the three areas: fetal monitoring, determination of uterine motility, and determination of oviductal motility. Let us review these three general areas and then look at a few other more specific areas where pressure measurements are being applied in reproductive biology research.

First, let us consider the fetal monitor. Figure 1 illustrates a clinical setup for a typical fetal monitor in an obstetrical unit using a commercially available monitor. There are four or five different major of these manufactureres instruments. Transducers are attached to the patient in various ways, and the monitor is used to record the activity of the uterus, the fetal heart rate, and in some cases the electrocardiogram of the fetus in the uterus before and during labor. The rationale behind this type of measurement is that the contractions of the uterine muscle, which are responsible for or at least contribute to the birth process, represent a physiologic stress placed on the fetus. This makes possible a kind of stress test in which we look at the fetal heart rate as a response to the uterine contraction. By observing how the fetal heart rate changes following uterine contractions, we are able to assess the physiological state of the fetus.

Thus, we see that this kind of monitor requires a means to obtain the fetal heart rate and to determine the activity of the uterus. In this chapter we are concerned with the latter problem. Uterine activity can be measured either externally (as illustrated on the patient in Figure 1) using tokography, or it can be done internally with the sensor directly coupled to the uterine cavity. First consider the external technique.

Figure 2 illustrates the tokodynamometer. Basically, it consists of a probe pressed against the abdominal wall. The probe depresses the fat layer, the abdominal muscles, and the myometrium, the muscle of the wall of the uterus. When the uterus contracts, it causes the anterior abdominal wall to feel stiff, just as when one contracts his biceps, he can feel his arm get firm. Basically, the tokodynamometric transducer senses the stiffening by means of the probe. The stiffened abdominal wall during uterine contractions tends to press the probe back into the body of the tokodynamometer, and a displacement transducer within the device senses this movement and generates an electrical signal proportional to it. Under ideal circumstances the tokodynamometer should not depress the abdomen but should only flatten it. In this way tension in the abdominal wall will not

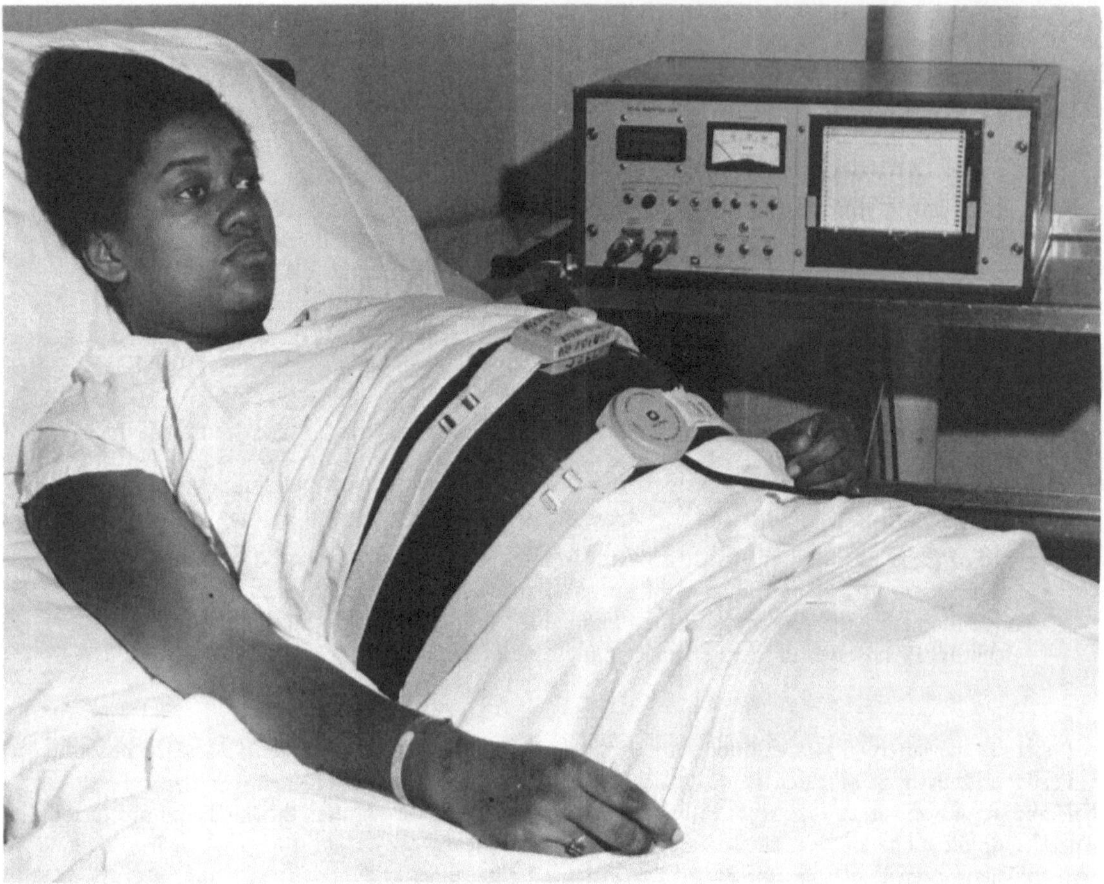

FIGURE 1. A commercial fetal monitor used to follow a patient in labor.

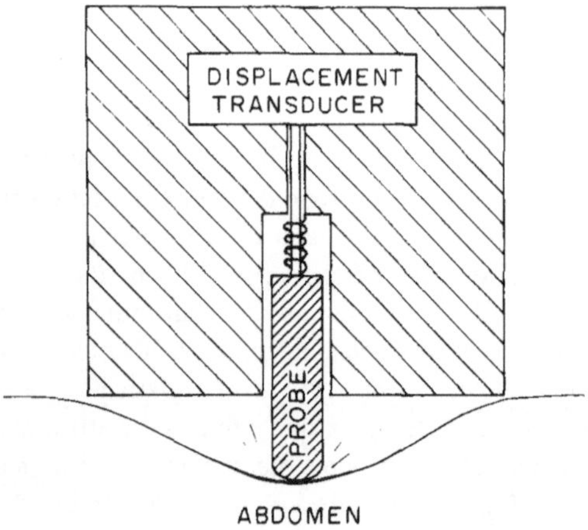


FIGURE 2. A tokodynamometer shown in cross-section, illustrating its principle of operation. The displacement sensor can be a strain gauge, a linear variable differential transformer, a variable resistance, or other type of displacement transducer. (From Roux, J. F., Neuman, M. R., and Goodlin, R. C., *CRC Crit. Rev. Bioeng.*, 2, 119, 1975.)

have a component normal to the transducer. Thus the probe will not respond to it, but pressure in the uterus will produce a force normal to the abdominal wall, and this will be transmitted directly through it to the probe of the transducer.

Although from this description one would imagine the tokodynamometer to be an adequate indirect pressure transducer, our experience has been that tokodynamometers generally are not very reliable and are seldom quantitative. We usually take the commercial instruments and modify them by increasing the length of the probe so that we depress the abdominal wall sufficiently to get some kind of recording. This we find helps us to detect contractions better, but in the process we have to completely forfeit the possibility of obtaining quantitative pressure data.

A similar technique which has been used by George LaCroix in Detroit is to have a fluid-filled dome structure, the base of which is an elastic membrane, intimately in contact with the abdominal wall and use this as the tokodynamometer.[1] The fluid in the dome can press against the abdominal wall and flatten it, allowing the intrauterine pressure to be transmitted to this fluid. The pressure in the fluid within this closed dome can then be measured by means of some type of pressure transducer. LaCroix has shown that there are cases where the pressure measured with this tokodynamometer agrees quite closely with pressure measured in the uterus using direct catheter techniques.[1] There are also cases where this good correlation is not observed.

We can see that tokodynamometers present many serious limitations as far as measuring pressure is concerned. However, if one wishes to monitor motility or to determine when uterine contractions are occurring and their relative strength, this device is more acceptable. There are nevertheless still several problems with the transducer. One is that it is sensitive to any kind of abdominal wall movement. Respiration, voluntary contraction of the abdominal muscles, and position changes of the patient all result in output from the transducer. As you can imagine, a woman in labor is not likely to lie still so that good recordings can be obtained, thus there is a great amount of artifact when recording patients in labor. Also, it is found quite often that it is necessary to reposition the tokodynamometer several times throughout the monitoring period. Many patients find these devices uncomfortable. Labor itself is uncomfortable and patients do not need the additional discomfort of having a fairly tight belt placed around their abdomen with a probe poking them in the abdominal wall.

A technique that has a much greater acceptance in terms of the data produced once it is possible to perform technically is the transcervical fluid-filled catheter coupled to an external pressure transducer, as illustrated in Figure 3. In this case, once the membranes surrounding the fetus have been ruptured, one can often introduce an open-ended catheter past the fetal head, or presenting part, to communicate with the amniotic fluid in the uterus. By coupling this to an external transducer it is possible to obtain an intrauterine pressure recording. This technique has been in use for about 20 years now, and it has been responsible for most of the physiological data which we have on the behavior of the uterus during labor and delivery.

There are also problems associated with this

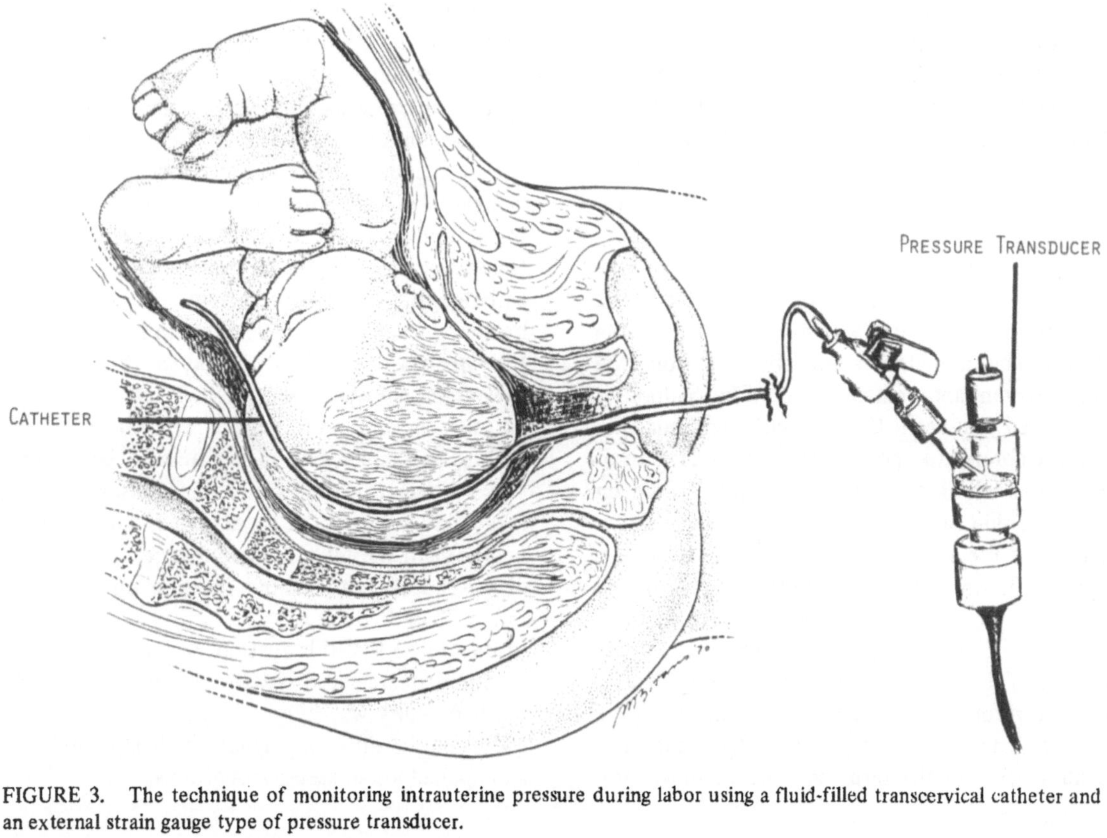


FIGURE 3. The technique of monitoring intrauterine pressure during labor using a fluid-filled transcervical catheter and an external strain gauge type of pressure transducer.

uterine pressure-measuring technique. First of all the catheter can very easily become obstructed. The uterus contains materials in addition to the amniotic fluid. There is a cheesy material that is known as *vernix caseosa* which can obstruct the tip of the catheter. Usually it can be blown off by flushing the catheter with normal saline. There is also a lot of tenacious mucous and the possibility of blood which can clot on the tip of the catheter. The catheter can also become dislodged from its position in the amniotic fluid, and its tip can become squeezed between a part of the fetus and the endometrium, the lining of the uterus; in effect it becomes uncoupled from the fluid and does not measure true intrauterine pressure.

Another problem is concerned with the catheter placement itself. Once the fetus is well engaged in the pelvis, it is difficult to introduce catheters into the uterus, and so in general placement must be done early in labor before the fetus has begun to descend into the pelvis. Yet one does not want to be too early and artificially rupture the fetal membranes, for the catheter can serve as a conduit for infection back into the uterus. One cannot leave a patient with ruptured membranes for periods greater than 24 hrs without running the risk of intrauterine infection.

The catheter system is sensitive to base line changes due to differences in elevation between the patient and the transducer. The technique described in the chapter by E. H. Wood and E. L. Ritman could be used to evaluate and compensate for this error.

Another technique that has been used for measuring intrauterine pressure on the pregnant uterus is the extraovular balloon technique developed by Csapo and co-workers.[2] They use a balloon-tipped catheter which is inserted into the uterus. The membranes are stripped from the endometrium and remain intact. The balloon is introduced between the endometrium and the fetal membranes to a point where it is in a region of the uterus where there is some amniotic fluid on the other side of the membranes, and the balloon can then sense pressure in this fluid. The balloon is filled with normal saline solution and coupled through the catheter to an appropriate transducer which then measures the pressure within the balloon. This technique has the advantage that a pressure recording can be made without rupturing the membranes, but obviously if the balloon is located between the fetus and the endometrium, the pressure in the balloon will result from forces between the fetus and the uterine wall rather than the amniotic fluid pressure. This technique also runs the risk that the fetal membranes may rupture upon introduction of the balloon.

One other technique that is used for measuring intrauterine pressure is the transabdominal catheter. A fluid-filled catheter is placed through a trocar through the abdominal wall, the uterus, and the fetal membranes. The amniotic sac remains essentially a closed system, since the membranes are punctured with a hole small enough just to admit the catheter, and the sides of the hole become effectively sealed against the endometrium. This method is for the most part not used in the United States because of the risk of infection. Also there is the possibility that the placement of the catheter can initiate uterine contractions and cause premature labor.

Our group has been working with indwelling pressure transducers to measure intrauterine pressure. Figure 4 illustrates the transmitter portion of a radiotelemetry system which utilizes a miniature radio transmitter that can be placed in the vagina and connected to a catheter-tip type of pressure transducer located in the intrauterine cavity. The advantage is, that some of the special problems listed above for the catheter system can be eliminated by using this type of intrauterine pressure sensors.

The intrauterine pressure sensor has some special requirements that one does not encounter with the catheter. The most important of course, is small size, since the transducer has to be easily introduced into the uterine cavity. The present transducer that our group is using is larger than the catheter that is used in conventional monitoring; however, when one considers it with respect to the tube that is necessary to introduce the catheter, they are of equivalent size.

Another important requirement for the intrauterine pressure sensor is that it must be insensitive to direct forces. Since the sensor will be located adjacent to the uterine wall (a contracting muscle) and the fetus (a moving object), it can get caught between them and measure direct forces rather than hydrostatic pressure. There must be protection so that this does not occur.

The pressure sensor must not harm the fetus or mother in any way either electrically or mechanically, and it must be capable of being sterilized and applied under aseptic conditions.

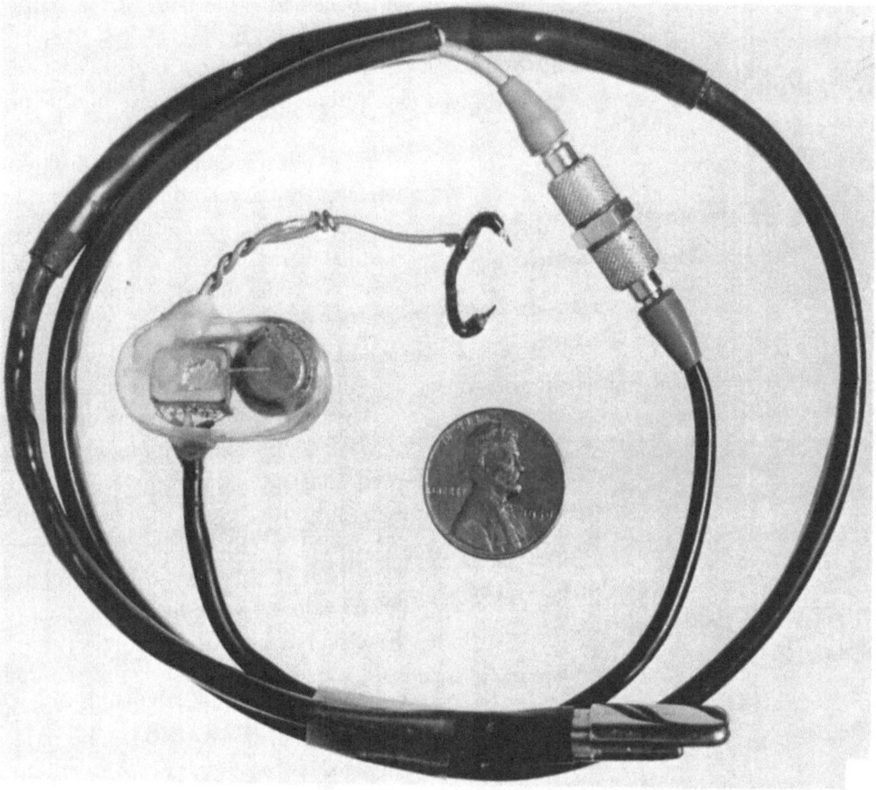

FIGURE 4. The transmitter portion of a fetal monitoring radiotelemetry system illustrating the intrauterine pressure sensor connected directly to the intravaginal transmitter unit. (From Neuman, M. R., Picconnatto, J., and Roux, J. F., *Gynecol. Invest.*, 1, 92, 1970. With permission.)

With these requirements in mind let us illustrate some of the transducers being developed at the Engineering Design Center at Case Western Reserve University. Figure 5A illustrates the use of the piezoresistive semiconductor strain gauge transducer. In this case the transducer is a Sensotec®* flat transducer that has been placed in a stainless steel shroud to protect it from direct forces. The entire front end of the structure is open so that a large communicating surface exists between the amniotic fluid and the transducer. There are additional ports on both flat faces of the structure, and these all help to minimize the risk of obstruction. The illustrated transducer was constructed with a conventional catheter on the back face so that comparative pressure measurements using the two techniques could be made.

Figure 5B illustrates a different type of structure. This uses a pressure-sensitive transistor (Pitran®**) type of transducer. This structure is circular in cross-section and has multiple openings in its tip so that the amniotic fluid can communicate with the pressure transducer. The tip can be removed exposing the pressure-sensitive surface of the transducer for cleaning. A similar but smaller structure, as shown in Figure 5C, makes use of the Millar*** catheter-tipped transducer. It is not possible to use this catheter-tipped transducer without the protection of the stainless steel shroud, because it would respond to direct forces as well as pressures. The diameter of this structure is approximately one half that of the shroud for the pressure-sensitive transistor.

In all three examples of the intrauterine pressure transducer, a small gauge hollow tube is included in the lead wire bundle. Upon introduction of the transducer, a stiff stainless steel wire stylet is placed in this tube to stiffen the structure so that it may be guided into the uterus by pushing it in from its proximal end. Once the

*Sensotec, Inc., 1400 Holly Avenue, Columbus, Ohio.

**Pitran, Stow Labs, Inc., 245G Kane Industrial Drive, Hudson, Massachusetts 01749

***Millar, Millar Instruments, Inc., P.O. Box 18227 (6001 Golf Freeway), Houston, Texas 77023.

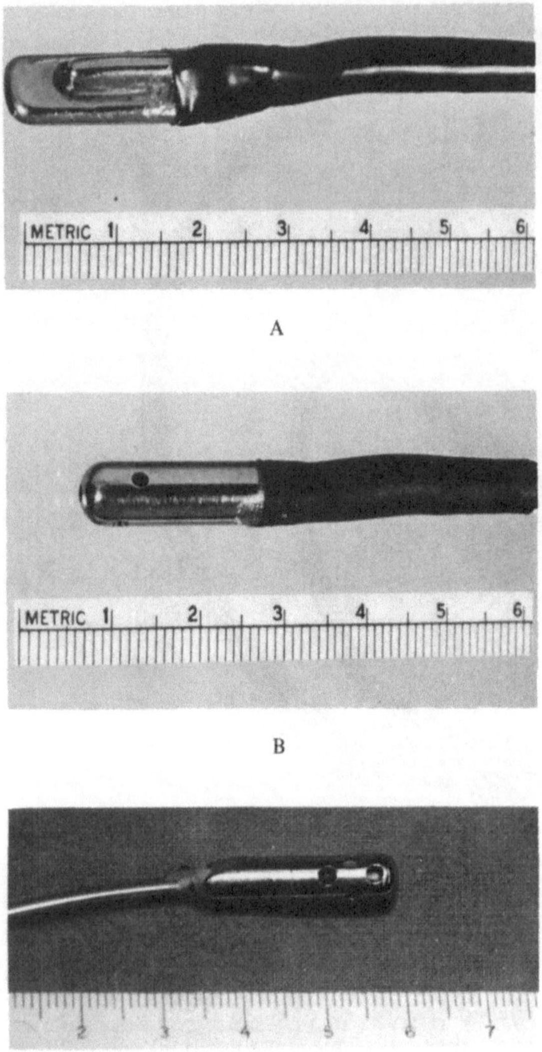


A

B

C

FIGURE 5. Various types of packaged intrauterine pressure sensors. A. Package for a flat aerospace-type silicon strain gauge transducer. B. Package for a pressure-sensitive transistor. C. Package for a 5F catheter-tip transducer. (From Neuman, M. R., Picconnatto, J., and Roux, J. F., *Gynecol. Invest.*, 1, 92, 1970. With permission.)

transducer is in place, the stylet is removed, and the cable assembly becomes quite flexible.

In comparing the pressure recordings from the intrauterine pressure transducer with recordings from an adjacent conventional catheter with external pressure transducer, we see from Figure 6 that the recordings show qualitative agreement. We shall see, however, that the agreement is not always quantitative and that errors introduced are not necessarily the fault of the transducers themselves.

Several years ago my colleagues and I were interested in comparing intrauterine pressure transducers, and we compared recordings of the same uterine contractions using different sensors.[3] Three pressure recordings were made from each uterine contraction, using either catheters and external transducers or internal intrauterine pressure transducers. Properties of the three pressure waveforms for a given contraction, such as amplitudes, waveshape, and duration, were determined and then compared.

When the amplitudes were considered, the ratio of peak amplitudes for pairs of transducers for given contractions were determined. Ratios rather than absolute values of contraction amplitudes were used so that calibration errors from one transducer to the next would be eliminated. If all the transducers were measuring the same pressure, the calculated ratios should always be unity or at least the same number from one contraction to another. It was found that this was not the case. For example, if we consider the ratio of pressure amplitudes for a set of contractions monitored during a given labor using two intrauterine catheters with external pressure sensors, we find the results illustrated in Figure 7. The means and standard deviations of this ratio are plotted as a function of the approximate cervical dilatation or progress of labor for a group of nine monitored labors. As can be seen, there is a great deal of variation in the ratio. As a matter of fact, the coefficient of variation of that ratio can become as large as 30 to 40%. Since the pressure amplitude ratio was determined using two identical pressure-sensing systems, we see that it will be difficult to compare different pressure-sensing systems, because the errors in the pressures being measured appear to be a result of the system in which they are measured rather than the transducers themselves. Thus, we see that there are severe limitations on the assessment of intrauterine pressure recordings during labor after membranes have been ruptured. Whether catheters or internal transducers are used, we find the same degree of discrepancy in the pressure amplitudes.

Whenever one plots a signal as a function of itself and uses the same scale on both axes, a straight line of unity slope results. If the two signals, however, differ from one another, the resulting plot will be something other than the line of unity slope. This technique can be used to

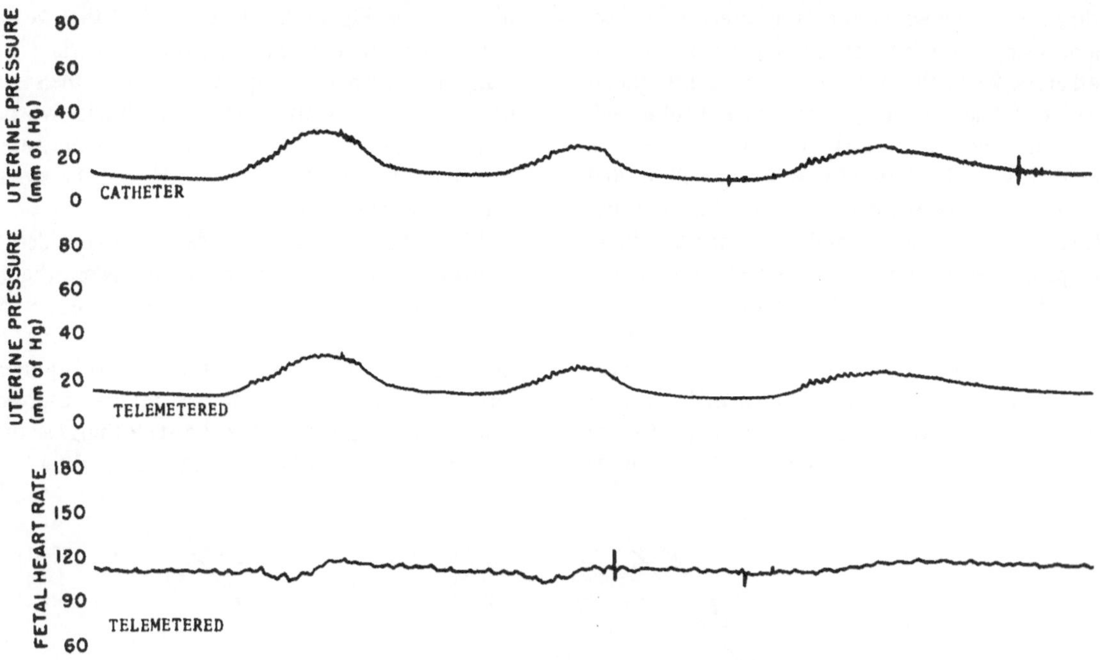


FIGURE 6. A comparison of intrauterine pressure data as determined by direct wire and radiotelemetry techniques. Top trace is a recording of intrauterine pressure monitored by a transcervical catheter with external strain gauge pressure transducer. Middle trace is the intrauterine pressure measured with an intrauterine sensor and transmitted by radiotelemetry. Bottom trace is the fetal heart rate determined simultaneously with the uterine pressures and transmitted by radiotelemetry. (From Neuman, M. R., Picconnatto, J., and Roux, J. F., *Gynecol. Invest.*, 1, 92, 1970. With permission.)

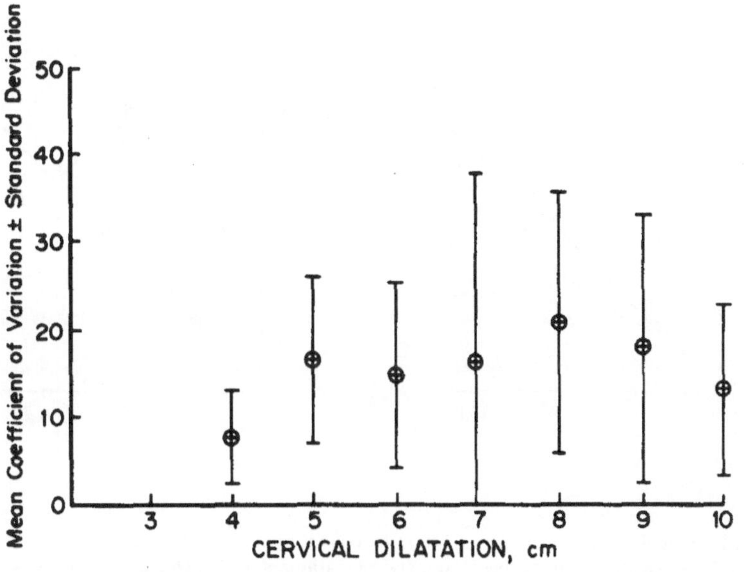


FIGURE 7. The coefficient of variance of the ratio of the pressure amplitudes of uterine contractions recorded simultaneously from two transcervical catheters with external strain gauge transducers, plotted as a rough function of cervical dilatation determined by the digital method and extrapolated on a labor graph. (From Neuman, M. R., Jordan, J. A., Roux, J. F., and Knoke, J. D., *Gynecol. Invest.*, 3, 165, 1972. With permission.)

compare recordings of the same uterine contraction, using different intrauterine transducers or catheters. When this is done, it is frequently found that the line of unity slope is not observed. Instead, curves such as those shown in Figure 8 frequently result. A typical recording is illustrated in Figure 8A for the case where the two pressure signals have the same general waveshape, but there is a proportional amplitude difference between the two of them. There can also be a situation where there is a nonlinear amplitude difference between the two recordings, or there can be a phase difference between them as shown in Figures 8B and C, respectively. More often than not we see combinations of all these types of distortion, as illustrated in Figure 8D. Thus, we find that we do not always measure the same pressures when we make measurements using two sensors in the same uterus. If this is the case, why should we be making transducers that measure uterine pressures to a high degree of accuracy if indeed the signal itself does not warrant such accuracy?

When the above described data were considered statistically, it was found that for a single catheter or pressure transducer placed in the uterus, a 95% confidence interval required a range of plus or minus 23% of the measured pressure. This puts a serious limitation on using the results observed from a single sensor to describe such things as the difference in effects between drug X and drug Y if

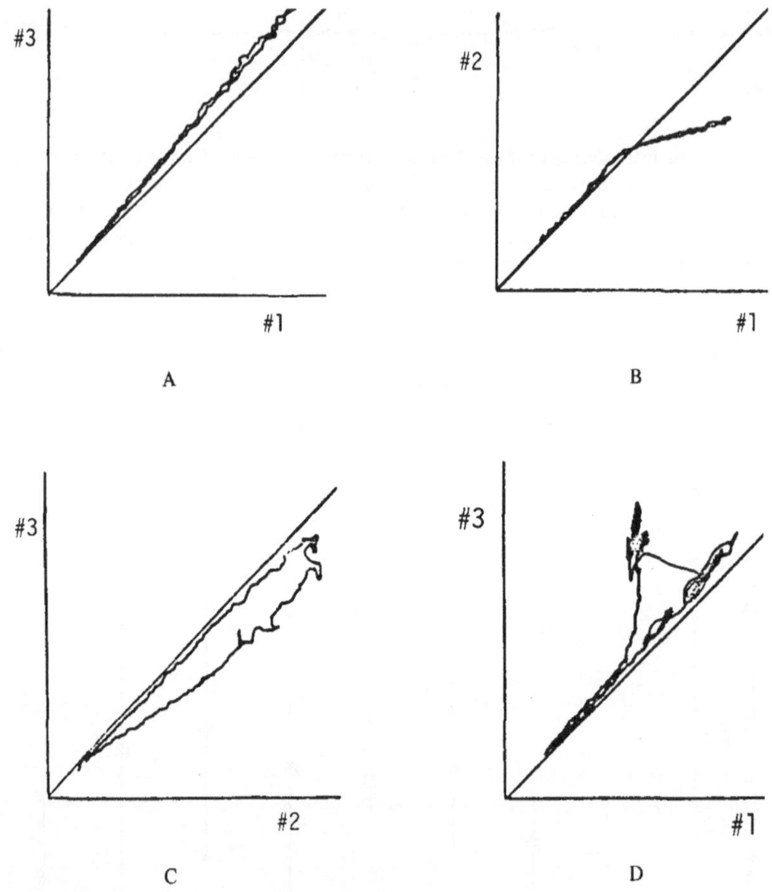


FIGURE 8. Four examples of simultaneous uterine pressure recordings during uterine contractions, using transcervical catheters and external strain gauge transducers. Each plot represents the signal from one transducer plotted as a function of the signal from the other. A. Difference in calibration constant between the two sensors. B. Nonlinear variations in the calibration constant. C. Phase distortion between the two sensors. D. Example of a typical recording with all of these combined.

that difference is relatively small compared to the confidence interval. The only way to handle this kind of a problem is to use more than one transducer or catheter and statistical analysis so as to compensate for the variability in the data.

There no doubt is a reason for this variation, but we can only speculate on this for the present. It is reasonable to assume that the two transducers or catheters will be located at different positions in the uterus and that Pascal's law does not apply because the uterus in labor is no longer a closed system. Furthermore, due to the fact that the uterus contains the fetus as well as the amniotic fluid and the fetal volume is much greater than that of the amniotic fluid present, it is quite possible that there will be a dynamic compartmentalization of the fluid, with separated compartments forming and joining up with other compartments as uterine contractions occur. The pressure within the compartments can easily differ while they are separated.

The important point is not the detail of what causes the variations in pressure signals, but rather than whenever a pressure transducer is to be used for a particular measurement it is very important that we understand the measurement and its limitations before we design a highly accurate measuring instrument. Quite often, as in the above example, such a design may be nothing more than an academic exercise due to basic limitations in the measurement itself.

Let us now consider some of the other applications of pressure measurements in obstetrics. These are currently used as research tools, and they are currently only being applied by a few investigators around the world. In studying the motility of the nonpregnant uterus, essentially the same techniques as described for the pregnant uterus are used. There are, however, additional considerations which must be made. Since the lumen of the nonpregnant uterus is for the most part a virtual space, there is not enough fluid within it to transmit pressure to a catheter using conventional techniques. Either one has to distort the anatomy by introducing some fluid into the uterus and blocking its exit out of the cervix, or one must measure pressure by slowly perfusing the catheter using an infusion pump at a constant flow. Either technique can result in errors.

Another technique is to use a balloon-tipped catheter instead of the open-ended type and to slightly inflate the balloon so that it begins to fill the potential space; thus contractions of the myometrium will cause an increase in the pressure within the balloon which is then transmitted through the fluid in the catheter to an external pressure transducer.

Recently some investigators have begun to consider the catheter-tipped transducer, such as that developed by Millar, for intrauterine pressure measurements in the nonpregnant uterus. There have, however, been problems with this device due to both direct mechanical forces against the transducer giving nonhydrostatic pressure recordings, and due to the stiffness of the transducer and its associated catheter irritating the myometrium and thereby increasing its activity.

There is indeed a great deal of information that could be gained from this type of measurement, but unfortunately, as was pointed out earlier in this volume, we really do not know what we mean by a pressure measurement in an organ such as this in the first place. Not only can we not understand the meaning of the pressure in the lumen of the uterus, but also in the uterine muscle (the myometrium) itself. In measuring the pressure within such a muscle, we no doubt would get anisotropic results, because the muscle fibers are not isotropically oriented. Some work has been done to measure myometrial pressures using microballoons injected into the myometrium, but the true meaning of such measurements remains obscure in view of the directional differences which would be lost. If a catheter-tipped sensor was placed in the muscle, would we measure pressure at all or would it be the strain in the transducer resulting from the muscular contraction?

The above comments concerned with the measurement of uterine pressures and uterine motility can only be amplified when we consider the motility of the Fallopian tube. The techniques for its measurement are generally the same; however, they must be scaled down even smaller since we are dealing with a much smaller anatomic structure. The techniques of open-ended, profused catheters and small balloons have been used in studying the motility of the Fallopian tube. In addition, external sensors have been placed on the tube to imply its internal pressure or at least to detect contractions. Miniature strain gauge type transducers have been placed on the outer wall of the Fallopian tube to measure its contraction, and there also have been transtubal measurements using magnetic fields, light transmission, and ultrasound. These techniques can all be used to obtain information about the pattern of activity in the

Fallopian tube, but none are able to get a true intraluminal pressure.[4]

An important consideration in making intrauterine and certainly intratubal pressure measurements is that anything we do to either of these structures will cause a distention of them which, since we are dealing with smooth muscle, can effect their behavior. Consider the gut as an example. If one distends the gut, one can easily alter its peristaltic activity, and hence, change its motility. It is reasonable to believe that any of these techniques described above will do the same to the Fallopian tube and possibly the uterus as well; thus size and irritability are important considerations in the development of pressure sensors for these structures.

Pressure measurements have been used in other areas of obstetrics research by various investigators, and three of these techniques will be briefly outlined. The pressure transmitted from the uterus to various locations within the fetus in pregnant animal models has been measured by using open-ended catheters introduced surgically into the fetus in utero and brought out through the uterus and maternal skin. The principal problem in such preparations is maintaining a patent catheter, as well as maintaining the transuterine and transcutaneous placement of the catheter in the awake animal.

Obstetricians and gynecologists are also interested in various abnormalities of the female urinary tract which can be the result of earlier pregnancy. A frequently encountered but not so well understood problem is stress incontinence, where increasing intra-abdominal pressure, such as occurs during a cough or laughing, causes the patient to involuntarily lose urine. Today, several investigators are examining the dynamics of the pressures in the female urinary tract in both normal women and those who suffer from this problem. Here it is important that the pressure sensor respond quickly and accurately, and that the miniature transducer which can be placed internally is clearly superior to catheter systems.

Another problem is that of studying the pressures necessary to open the uterine cervix. Some patients suffer from a so-called incompetent cervix which is responsible for late miscarriages in pregnancy. While this problem is known to exist, there currently is no good diagnostic procedure for detecting it. Recent work by the author and co-workers has been directed towards developing a balloon which can be placed within the cervical canal and inflated using sterile water. A pressure-volume characteristic of the balloon during this inflation indicated, in a preliminary study, that for those patients with historical evidence of incompetent cervix, the balloon could be filled at a much lower pressure than is required for those patients who are considered normal.

In conclusion, there are three basic, current considerations associated with pressure measurements in the reproductive system:

1. We must find methods to improve the reliability of our measurements.
2. It is important in physiologic and clinical studies that we have transducers of the smallest possible size for ease of introduction and minimal interference with the tissues being studied.
3. It would be far better if noninvasive or microinvasive techniques could be developed for such pressure measurements.

## DIALOGUE

**Dr. McCutcheon**: I wonder if you can detect uterine contractions by changes in anterior abdominal wall shape? I was also wondering if ultrasonic echo transducers could be included in the tip of your displacement measuring device. Why do you not use abdominal wall bearing instead of trying to get an estimate of the chamber pressure?

**Dr. Neuman**: The echo has not to my knowledge been done, but it should work, although it might be a bit complicated. The Whitney gauge, the mercury in an elastic tube strain gauge, has been put around the abdomen, but I cannot remember by whom. You can see a contraction, and the abdomen changes shape so this could be recorded. The problem is that the patient does not like to have her abdomen restrained in any way, and she does not even like the belt on the tokodynamometer because it must exert some force to hold the device in place. It has been our experience and probably that of others as well that patients, when they have these exterior monitors, always complain about the straps.

**Mr. Konigsberg**: Is is possible (this is speculation) that they are complaining about the straps in terms of physical discomfort, or is it the fact that they are wired up, the fear element of being wired up affecting the patients? I have heard that the latter is a definite possibility.

**Dr. Neuman**: I do not have any data but I

would guess it is. I have usually found the patients who are most upset and anxious about their labor are the ones that complain the most.

**Dr. Geddes**: They prefer to be wired up with the wires, but without the straps.

**Dr. Neuman**: Right, I think that is one of the things that led us to the telemetry, because then the patient, after the instrumentation procedure, can forget that she is being monitored. We do not usually leave the receiver in the same room with her. There is no need for her to see the data. As a matter of fact, we had one case where we had to remove the receiver from the room because the husband was more interested in the data than in his wife.

**Dr. Geddes**: I notice that you had fetal EKG electrodes on the head. Where are the others?

**Dr. Neuman**: The fetal EKGs are a unipolar type of system. The electrodes are actually measuring the transcutaneous potential. The electrodes consist of a fine wire that penetrates the skin and the reference electrode is just a loop of wire on the back of it in the amniotic fluid.

**Dr. Geddes**: I was wondering if you could get a catheter with a metal tip and use it as an electrode?

**Dr. Neuman**: I have not myself, but this has been done. It is not as reliable as a direct connection to the fetus, but it is better than having the electrodes on the abdomen.

**Dr. Geddes**: I mean in addition to the scalp electrodes?

**Dr. Neuman**: You mean as reference electrodes?

**Dr. Geddes**: Using the example of the catheter-tipped electrodes?

**Dr. Neuman**: We have not done it, it would be interesting.

## REFERENCES


1. **LaCroix, G. E.,** Monitoring labor by an external tokodynamometer, *Am. J. Obstet. Gynecol.*, 101, 111, 1968.
2. **Csapo, A.,** the diagnostic significance of the intra-uterine pressure, *Obstet. Gynecol. Surv.*, Part I, 25(5), 403; Part II, 25(6), 515, 1970.
3. **Neuman, M. R., Jordan, J. A., Roux, J. F., and Knoke, J. D.,** Validity of intra-uterine pressure measurements with transcervical intraamniotic catheters and an intraamniotic miniature pressure transducer during labor, *Gynecol. Invest.*, 3, 165, 1972.
4. **Blandau, R. J., Boling, J. L., Halber, S., and Verdugo, P.,** Methods for studying oviductal physiology, *Gynecol. Invest.*, 6, 123, 1975.

# CLINICAL ASPECTS OF CARDIOVASCULAR PRESSURE MEASUREMENTS IN PEDIATRICS

**Gordon Borkat**

Cardiovascular uses of indwelling pressure transducers in pediatrics are of most value in neonatal, surgical, and medical intensive care units, and during cardiac catheterization.

Newborn intensive care units employ umbilical artery catheters for intermittent blood pressure measurements. Indwelling pressure transducers would facilitate accurate continuous evaluation and aid in the diagnosis and management of pneumothorax, patent ductus arteriosus, ductal-dependent systemic flow, and therapy for hypovolemic shock.

General intensive care units measure arterial, pulmonary arterial wedge, and central venous pressures in patients with shock, and in postoperative cardiothoracic surgical patients. Continuous monitoring is essential to balance fluid input versus output and to help choose specific pharmacologic interventions.

During cardiac catheterization, indwelling pressure transducers, with echocardiography, have become invaluable for clinical research regarding hemodynamic events and myocardial contractility.

From the clinical standpoint of need, the following characteristics of indwelling pressure transducers must be considered: facility of application, size, flexibility, adaptability to use with other physiologic measurement modalities, thrombogenicity, durability, reliability, and general safety.

The practice of clinical pediatrics has measured intravascular and intracardiac pressures for many years. In most cases this has been accomplished by traditional fluid-filled catheters connected to external strain gauge transducers. More recently, indwelling pressure transducers have occasionally been used for clinical research.

This chapter will focus on those areas of pediatrics which utilize intravascular and intracardiac pressure measurement. Additionally, those characteristics of intravascular catheters and indwelling pressure transducers which are of importance to the clinical pediatrician and pediatric cardiologist will be mentioned.

By far, the most important application of intravascular catheters is in the newborn intensive care unit. The patients admitted to the unit at Rainbow Babies and Childrens Hospital in Cleveland, Ohio are premature infants, or newborns, who are considered to be at high risk because of difficulty during pregnancy or delivery. Either during transportation from other hospitals, or on admission to the intensive care unit, a catheter is placed in the umbilical artery and advanced so that its tip is in the abdominal aorta. The presently used catheters are polyethylene, of very small size (with internal diameter 0.023 in.), and are fluid-filled from the indwelling tip to the external connection. With this catheter in place, measurements of blood pressure are made, usually intermittently. Additionally, samples of blood can be withdrawn for determinations of gases and chemistries, and fluids and medications may be administered.

Continuous measurement of blood pressure, rather than intermittent, would be invaluable in assessing adequacy of therapy and in diagnosing serious problems encountered in the newborn infant. During delivery, some newborns may have significant blood loss, either from abnormal separation of the placenta or transfusion to a twin. The resulting hypovolemia and hypotension must be quickly evaluated and treated. Since the amount of blood loss is unknown, the only accurate means of determining efficacy of replacement therapy is by continuous blood pressure measurement.

Premature newborns with respiratory distress or aspiration occasionally develop a pneumothorax, causing further respiratory embarrassment. Since the blood pressure decreases immediately, recognition and follow-up with a chest X-ray at that time easily gives the diagnosis.

Cardiovascular abnormalities can also be recognized by indwelling pressure transducers. An increase in the pulse pressure is found with the diagnosis of a patent ductus arteriosus with blood flow from the aorta to the pulmonary artery. In the patient who also has respiratory distress, this can be associated with clinical deterioration. Finally, some infants have cardiac abnormalities where the major blood supply to the body is via a patent ductus arteriosus. There may be obstruction on the systemic side of the heart, such that flow goes by way of the pulmonary artery across the ductus to the body. Changing pressures can indicate a dynamic state and predict which infants should undergo diagnostic cardiac catheterization and possible surgery.

In newborn intensive care units, indwelling catheters can be in place for many days and have numerous problems. The blood pressure tracing is often damped, which requires repeated flushing of the catheter with a heparinized solution. This may fluid-overload a small premature infant. Additionally, there may be thrombus formation and subsequent occlusion of a vessel, or embolization to the lower extremities after flushing. There may also be introduction of air or foreign materials. Finally, because catheters are potentially open from the patient to the environment they have inadvertently been disconnected, causing severe and occasionally fatal hemorrhage.

At the present time, advances are being made to develop indwelling electrodes for the measurement of blood gases. Combined with indwelling pressure transducers, an entirely closed system would be available, reducing some of the risk, and allowing continuous high-fidelity pressure recording.

Diagnostic pediatric cardiac catheterization is usually a procedure lasting up to 3 hr. During that time, extensive physiologic data, including blood pressure measurements and oxygen saturations, are obtained, and injections of radiopaque contrast material are made into selected chambers of the heart and great vessels.

During routine studies, blood pressure is measured by fluid-filled catheters connected to external strain gauge transducers. In most cases this has proven to be a reliable method. However, some infants have small blood vessels which only allow passage of 4F catheters, approximately 1.5 mm in diameter. Blood pressure measurement is barely adequate, with severe damping, and the catheter requires frequent flushing (Figure 1). Since these patients have heart disease and may be in congestive heart failure, fluid overloading is a serious problem. Also, dislodged thrombi may by-pass the

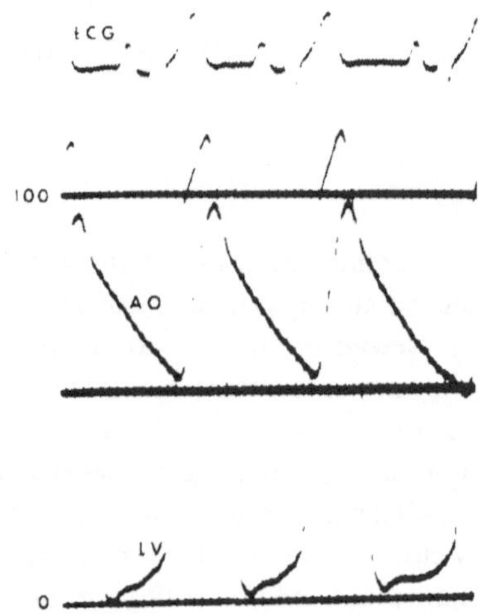


FIGURE 1. Simultaneous pressure recordings from left ventricle (LV) with large diameter fluid-filled catheter and aorta with 4F fluid-filled catheter. Aortic pressure (AO) is markedly damped, revealing apparent aortic valve gradient.

lungs and embolize to the systemic circuit, including the brain.

While catheter-tip pressure transducers have been of some clinical value in older children, because of stiffness they may be dangerous in small infants. The newborn heart is approximately 4 cm in its greatest dimension, and the wall is only a few millimeters thick. This makes the risk of perforation extremely high.

There are two major uses of indwelling pressure transducers during clinical research in pediatric cardiac catheterization. To explain the hemodynamic and physical events of the cardiac cycle, high-fidelity pressure recordings without artefact are essential. Simultaneous recordings of pressures in two areas can explain such complex phenomena as bidirectional flow across atrial and ventricular septal defects in transposition of the great arteries. Pressures can also be compared with intracavitary phonocardiography and echocardiography to explain the genesis of heart sounds and murmurs. The laboratory at Rainbow Babies and Childrens Hospital has recently begun to measure left ventricular pressure with echocardiography and phonocardiography to determine the relationship of the closure of the mitral valve to the first heart sound. When such relationships are better understood, the evaluation of the abnormal cardiac state, using noninvasive techniques, will be more accurate.

The major unanswered question in pediatric

cardiology is the long-term effect of disease states with volume and/or pressure overload on the myocardium. The timing of surgical correction must be based not only on surgical technique, but also irreversibility of myocardial damage. At present, one method for assessment involves indwelling pressure transducers for ventricular pressure measurements and derivatives (Figure 2). For determinations of myocardial contractility in population studies, this technique has not been helpful. However, serial data on the same patient have helped assess the state of the diseased left ventricle. Further work is in progress evaluating the right ventricle. Combined volume, flow, and pressure measurements will probably be most helpful.

The last major area in pediatrics where blood pressure measurements are crucial is in intensive care units for postoperative cardiothoracic surgery patients. In most institutions, management of these patients is a joint effort by cardiovascular surgeons and pediatric cardiologists. Fluid-filled catheters are placed to monitor central venous pressure, left atrial pressure, and systemic arterial pressure. Evaluation of the hemodynamics of decreased blood pressure is better understood, and transfusions or vasopressors can be given. Increased left atrial pressure suggests cardiac failure, so that inotropic cardiac agents should be administered. Finally, in view of myocardial failure unresponsive to routine therapy, systemic vasodilators, such as nitroprusside, might be given to decrease afterload on the heart. Again, careful titration against continuous blood pressure measuring is essential.

From the viewpoint of the pediatrician or pediatric cardiologist involved with clinical medicine, indwelling pressure transducers must have characteristics beyond those generally demanded in devices used for measuring other physiologic parameters.

Obviously, they must have an appropriate frequency response, thermostability, reliability, accuracy at different levels of pressure measurement, and simple calibration and zero reference.

In addition, pediatricians require small transducers, which can readily be introduced into 1.5-mm vessels with minimal trauma (Figure 3). They must have tips which are soft, minimizing the possibility of vessel or cardiac perforation. Moderate flexibility is essential to insure maneuverability through tortuous vessels and through the heart. Recently, very flexible balloon-tipped, flow-directed catheters which are used with external strain gauge pressure transducers have become available. They would approximate our needs if combined with indwelling pressure transducers and oximetry.

The major problem with intravascular devices is thrombus formation and its occasional dislodgment with embolization. During the few hours of a cardiac catheterization, patients receive large amounts of anticoagulant, preventing major thrombi from occurring. In seriously ill infants who require surgery, an indwelling catheter for central venous pressure measurement and fluid administration is placed, and there is minimal further heparinization. Large venous thrombi

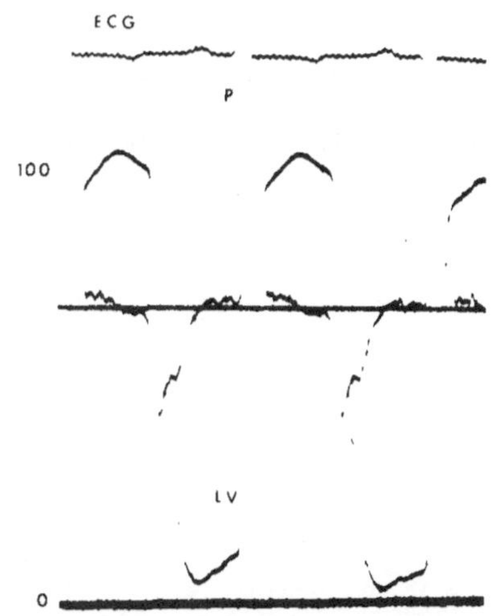


FIGURE 2. Left ventricular (LV) pressure tracing with derivative (dP/dt).

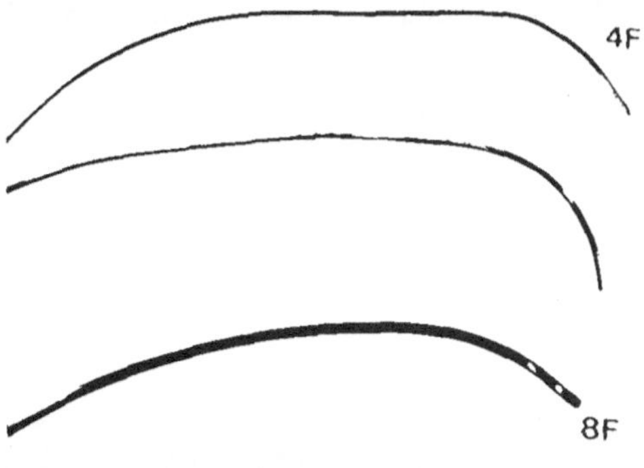


FIGURE 3. Range of catheter sizes used in pediatrics, including 5F Millar Mikro-Tip® catheter pressure transducer (middle).

form, occasionally occluding the inferior vena cava, with venous return via a paravertebral plexus. Umbilical venous catheterization in newborn intensive care units often caused this complication, so that the procedure is now rarely performed. Thrombi, with or without embolization, can also occur on the arterial side, with loss of pulses, occasional necrosis of digits, and subsequent amputation.

Because of manipulation, it is crucial that indwelling pressure transducers have high tolerance to the stress of sharp bending and lengthwise torque. There should be low risk of catheter or wire fracture. Coupled to fiberoptic oximetry, which is presently being evaluated, there would be reduced possibility of embolization and air and foreign-body introduction. For our purposes, indwelling pressure transducers should be easily cleaned and sterilized, and be reusable. Finally, they must be electrically safe.

In conclusion, the actual use of indwelling pressure transducers in clinical pediatrics and pediatric cardiology has been minimal because of limitations in the devices available. The present application during cardiac catheterization has been quite valuable, and potential for future use is enormous.

## DIALOGUE

**Dr. Kahn**: The comment was made about catheters in the vascular compartment being electrically hazardous. Could you qualify that?

**Dr. Borkat**: I think that catheters should be very carefully insulated, because the amount of current which may cause ventricular fibrillation is substantially less when the tip is in or very close to the heart. In other words, the currents required for micro shocks are much less than those needed for macro shocks.

**Dr. Kahn**: Yes, but doesn't that tip have to be within a millimeter or two of cardiac muscle to do that?

**Dr. Borkat**: It is true that we may satisfy those conditions with cardiac catheterization. In addition, it is also very easy to bury one of these devices in the ventricular wall and thereby damage the transducer – particularly when one introduces the transducer into the chamber and pulls it back out once it is lost.

**Dr. Geddes**: It should be pointed out that there is an additional shock hazard when one uses intracardiac catheters along with electrosurgical devices.

## REFERENCES


1. **Kitterman, J. A., Phibbs, R. H., and Tooley, W. H.,** Catheterization of umbilical vessels in newborn infants, *Pediatr. Clin. North Am.,* 17, 895, 1970.
2. **Stanger, P., Heymann, M. A., Tarnoff, H., Hoffman, J. I. E., and Rudolph, A. M.,** Complications of cardiac catheterization of neonates, infants, and children, *Circulation,* 50, 595, 1974.

# USES IN THE GENITOURINARY AREA

**Gerald W. Timm**

Indwelling pressure transducers find application in three types of measurements in the genitourinary area: (1) as a means of measuring pressure in the bladder to indicate volume; (2) as an indicator of effectiveness of bladder stimulation for urinary evacuation; and (3) as a means of measuring pressure in inflatable devices for urethral occlusion. The experimental use of these techniques is described in this presentation.

Many studies involving the measurement of pressures in genitourinary function and attempts to find control systems to regulate these pressures have been conducted in the last 10 to 15 years.[1] A portion of this research and examples of other applications of pressure control and regulation techniques will be discussed in this chapter.

The function of the lower urinary tract is primarily storage with intermittent expulsion of urine. Its cyclical function is similar to the cardiac system; however, the micturition reflex is a socially conditioned reflex. This means that patient control is needed. Patient control is lost due to various neurological diseases or injury to the nervous system.

The resulting loss of the flush of the urine, the millpond effect with the pooling of the urine, causes infection. So, it becomes important that the patient is able to maintain this constant flushing action. Pressurization and pressure-monitoring devices are basic for three purposes in the lower urinary tract.

Figure 1 shows the first area in which pressure monitoring is used as a means to measure bladder volume. In a given patient, basically the pressure in the bladder is found to increase in a fairly reproducible fashion as a function of the volume. Unfortunately, this pressure is not sufficiently dependent upon the volume to produce a reliable measurement.

Secondly, work has been done with indwelling pressure transducers which are placed directly into the bladder. Dr. Donald Gleason of the University of Arizona was the first to report the use of the transducers.[2] He found that the pressure capsules frequently led to calcification and formation of bladder stones, and they ultimately resulted in bladder infections as a result of foreign body presence.

Our experience had been similar to his. We have not found any materials that have been compatible with urine and that have not produced calcification and stones. What we have worked on since is a technique for measuring bladder volumes by a means to measure bladder wall distension.[3] This is not a pressure-measuring technique, but a proximity measurement between two capsules that are placed on the wall of the bladder. Basically, a magnet is placed on one position and a reed switch with telemetry capsule is placed at another known

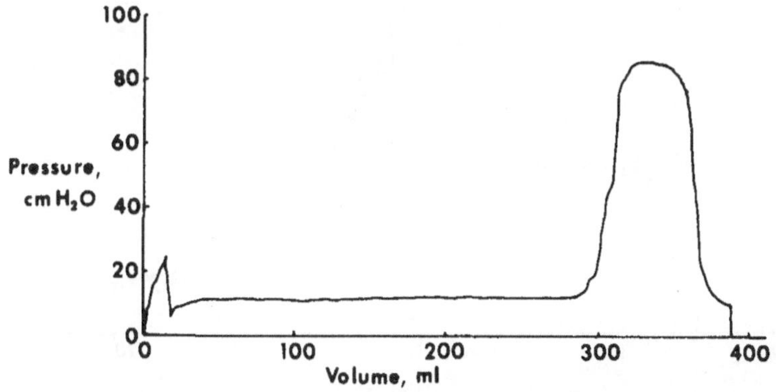


FIGURE 1. Pressure developed in the urinary bladder as a function of volume.

position. The reed switch turns on the telemetry capsule when the bladder volume exceeds the predetermined amount. The bladder volume sensor ceases telemetering when the volume of urine falls below this threshold volume due to urine evacuation.

In other words we fix the trigger point at some known calibrated volume and it will begin telemetering a signal whenever the contents of the bladder exceed this number. It could be set for 400 ml of urine or whatever the surgeon predetermines. The device has not yet been implanted in humans but is simply an animal experimental device.

These pressure-sensing devices are used in conjunction with electrical bladder stimulation to produce bladder evacuation.[4] With an indwelling pressure telemetry capsule, one could determine how effective the electrical bladder stimulator was in producing detrusor muscle contraction. However, this again has the same problems mentioned previously – the capsules produce bladder stones.

With bladder stimulation, a rapid rise in bladder pressure occurs following stimulus onset (Figure 2). However with long-term implantation with the electrodes in the wall of the bladder, the response curve decreases, and consequently the intensity of stimulation must be increased to obtain the equivalent response. At the present time it is not known how much the intensity should be increased. Bladder pressure sensing and telemetry would provide the information needed to assess effectiveness of stimulation.

The third area where bladder and urinary control devices have been developed is in urethral occlusion to prevent urinary leakage.[5,6] Thus, it is a treatment for urinary incontinence (Figure 3).

What it consists of is an inflatable cuff much like a blood pressure cuff, but made of implantable grade materials and positioned around the urethra at the bladder neck. The pressure in the cuff is controlled by a series of valves and pumps so that a certain maximum pressure cannot be exceeded.

What we have found in our experimental animals is that a pressure of 70 to 90 cm of water was well tolerated by the urethral tissue. About 150 cm of water was necessary before the blood flow in the tissue could be reduced by 50%. Thus, setting the pressure in the range of 70 to 90 cm of water provides a good safety margin.

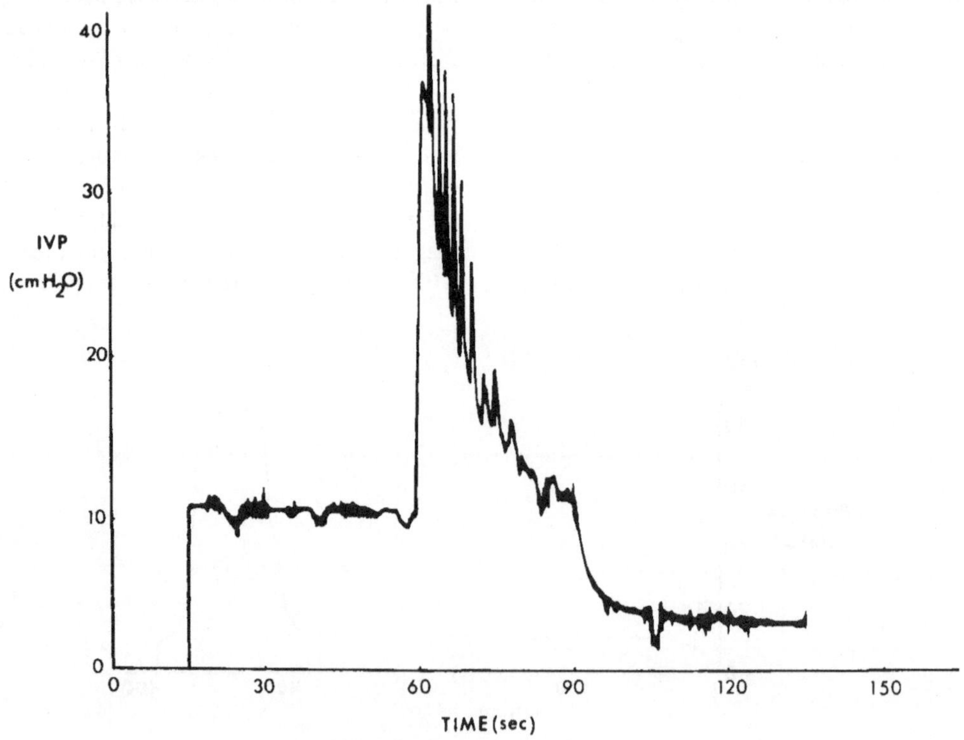


FIGURE 2. Pressure response of bladder to electrical stimulation of the detrusor muscle.

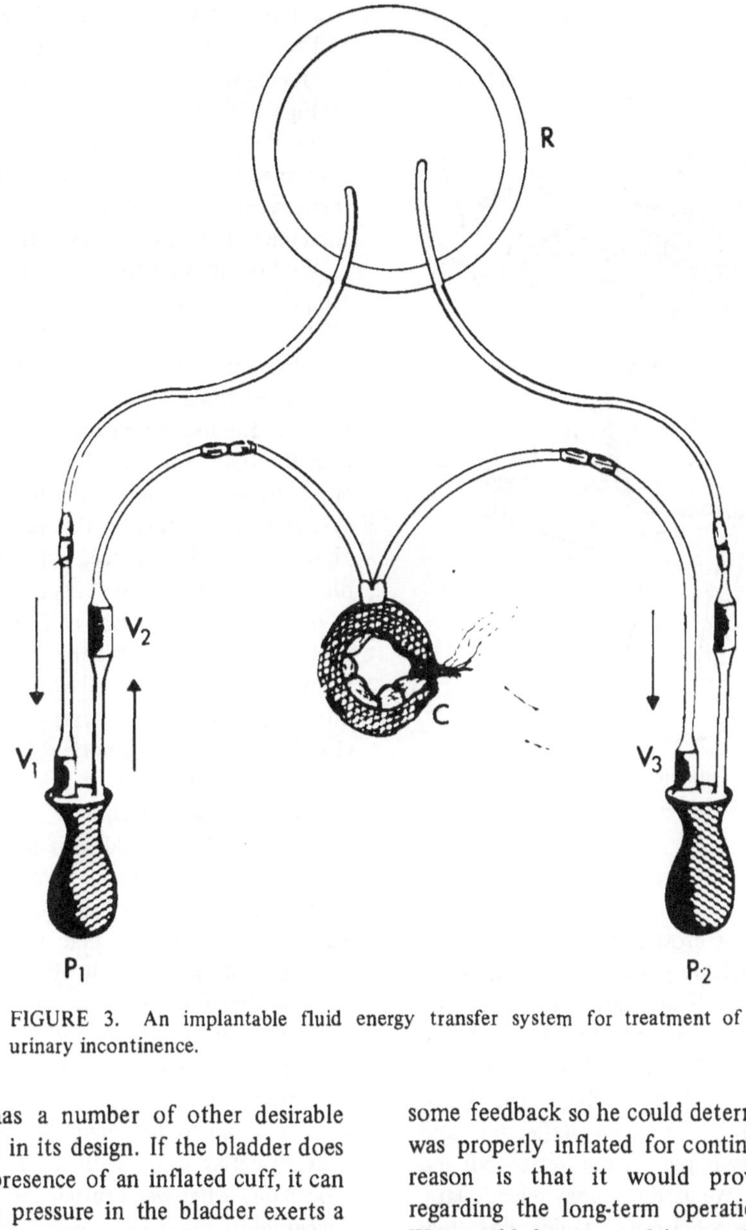


FIGURE 3. An implantable fluid energy transfer system for treatment of urinary incontinence.

The device has a number of other desirable features inherent in its design. If the bladder does contract in the presence of an inflated cuff, it can be emptied. The pressure in the bladder exerts a tissue pressure against the cuff, which forces fluid from the cuff back into a reservoir positioned beneath the abdominal muscle, and urine leakage can then occur past the partially deflated cuff.

Leakage does not occur when one coughs or strains as the reservoir pressure balances the cuff pressure so that leakage does not occur (Figure 4). Thus, the cuff will not deflate under those conditions.

The development of a pressure telemetry system for use with urinary occlusive devices would be desirable for two reasons. One is that it would serve as a useful technique for patient-training purposes. This then would give the patient some feedback so he could determine that the cuff was properly inflated for continence. The second reason is that it would provide information regarding the long-term operation of the device. We would be sure of having information on a continuous basis to insure that the characteristics of the valves have not changed with time. This data would help to determine the minimum pressure in the cuff required to produce urinary continence.

One of the very important considerations of the proposed pressure telemetry capsule was the cost factor. The device should have the capability of being incorporated into a system that could be used routinely in patients.

The pressure telemetry capsule would be used as shown in Figure 5. We would monitor the pressure in the cuff and telemeter that information

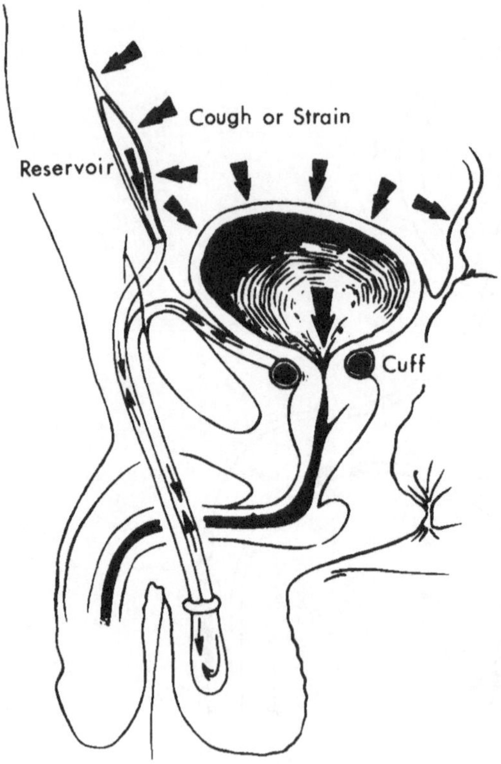


FIGURE 4. Pressure equilibration in the artificial urinary sphincter during coughing or straining.

outside the body by means of telemetry techniques.

Figure 6 shows a cross-sectional view of the transducer used. It is very simple in its design; a Colpitts oscillator is being used. A is the inlet tubes which are connected to the cuff; B represents the coil; C is the electronics; D is the Teflon® form; E is a Silastic® diaphragm which was used to give a maximum deflection and maximum sensitivity in moving the ferrite rod; and F is the ferrite rod. We used an epoxy encapsulate, G; H is the pressure chamber which was then sealed, and I represents a sealed air chamber. The sensitivity of the entire device was found to be 2.48 kHz/cm $H_2O$ when it was implanted in an experimental animal.[7]

There was a huge shift in the base-line frequency following implantation (Figure 7). It was initially about 462 kHz, and it increased to a center frequency of 467 kHz after about 4 months. Again, the sensitivity is about 2.4 kHz/cm $H_2O$. Since we are trying to measure pressures in the range of 50 to 100 cm $H_2O$, this shift is of great concern to us. This overall frequency increase corresponds to a change of 1.4% of the base-line frequency. We, therefore, did some additional studies to determine exactly what was producing this shift in the base-line frequency.

Figure 8 shows our experimental setup to check the pressure telemetry capsule. It consists of a temperature control bath in which we would immerse the capsule. We could also control the fluid environment of the capsule.

We first allowed the device to stabilize at room temperature, then increased the temperature to 37°C, and it stabilized at point A (Figure 9). The base line frequency began at 107 kHz in this case, and it increased to 108.7 kHz. We believed that this was due to the temperature dependence of the coil, the transistor, and the capacitor which cause an increase in the oscillating frequency. The system was then filled with a radiopaque dye, 25% Hypaque.® When placed in contact with the diaphragm within the telemetry capsule, the frequency went up an additional amount (to position B). We felt that this was due to diffusion of water through the Silastic membrane, which produced certain vapor pressures within the sealed chamber due to the osmotic diffusion of the water across the membrane.

The first two parts were conducted with the capsule itself being inside of a plastic bag so that it could not be in contact with the water in the temperature bath. The bag was then placed into a water bath. We felt that the final stabilized frequency shown after 30 days was then due to the additional changes due to the diffusion of water through the epoxy and Silastic coatings. We were not sure how to explain the effects, except that probably it changed the resistor values. We were not able to document this resistance change after dissecting some of the devices. Basically, this is probably what was causing the pressure shift in the capsule.

Some other applications for pressure measurements in the urinary tract should also be discussed at this point. One of the things that we are concerned with is the distribution of pressure beneath the inflated cuff. Knowledge of this distribution would help if there are any particular points between the cuff and the urethra, where we might get localized high pressure. However, research in this area has not been successful.

A method such as that shown in Figure 10 could measure the distribution of forces between a soft fiber cuff and softer, more pliable tissue, but would not interfere with the interaction between

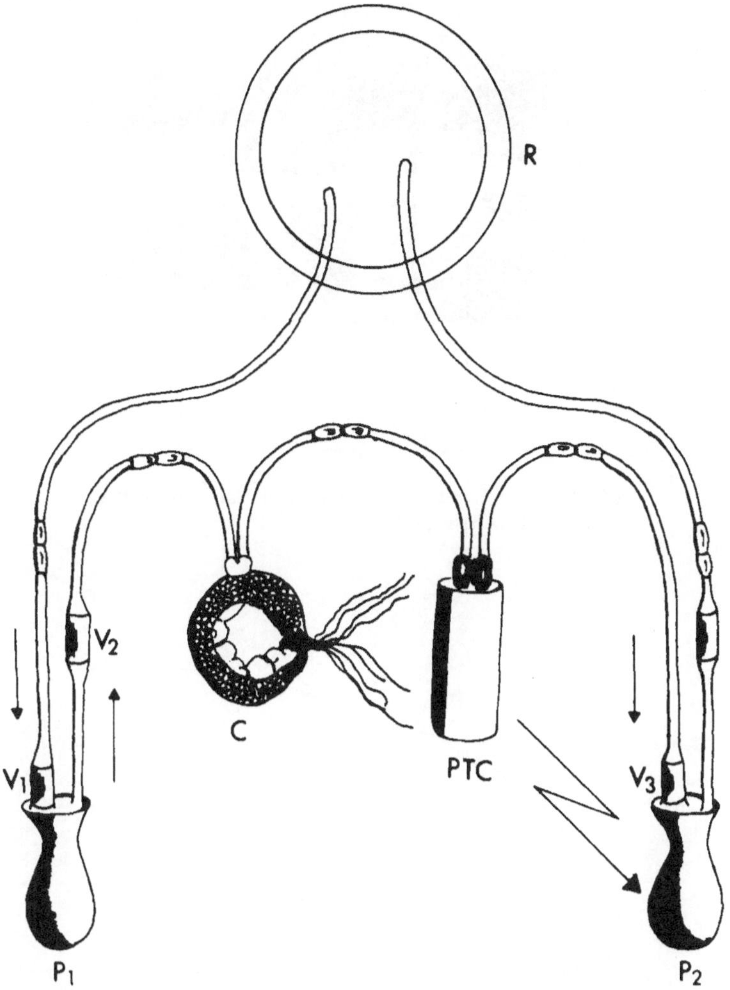


FIGURE 5. Pressure telemetry capsule attached to the artificial sphincter system to monitor cuff pressure.

the cuff and the urethra, would be very valuable.

Figure 11 demonstrates the results of a technique for measuring urethral pressure profiles inside the urethra and beneath the cuff for various pressures.[8,9] Curve A is for the deflated cuff. A catheter is inserted which has a port on the side of it through which water is infused. At the point of the arrow, the water no longer flows outside of the urethra, indicating that it is probably the cross-over point, that is, the water then presumably goes into the bladder.[10]

Curve B represents an inflated cuff with 50 cm $H_2O$, curve C with 75 cm $H_2O$, and curve D with 100 cm $H_2O$. There is good correlation between the pressure in the urethral mucosa and that in the cuff.

Another use for the implantable occlusive technique is for the regulation gastrointestinal flow.[11] This has application in ileostomy or colostomy patients. We have done some experimental work on this in animals, but it has not been done clinically at this point. Some of the potential applications in this area are still being investigated.

Another area where this implantable fluid energy transfer system has been used is for treatment of penile erectile impotence in the male.[12] In this case, the inflatable cuff is replaced with two inflatable cylinders which are then inserted into the corpora bodies of the male and can be inflated and deflated by the pumping system. This has been used clinically in a number of patients.[13]

In conclusion then, pressure telemetry tech-

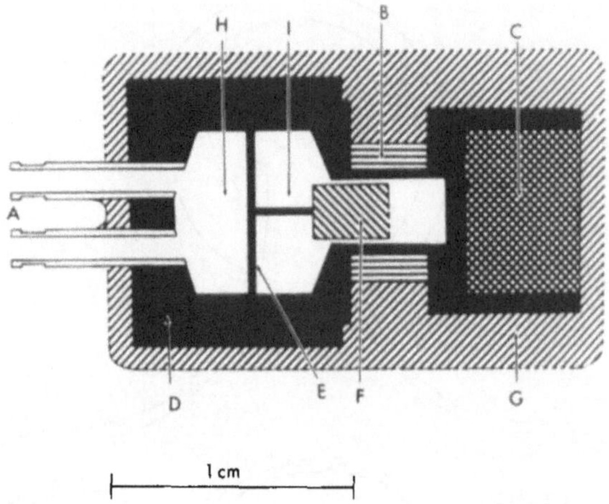


FIGURE 6. Construction drawing of an implantable pressure telemetry capsule.

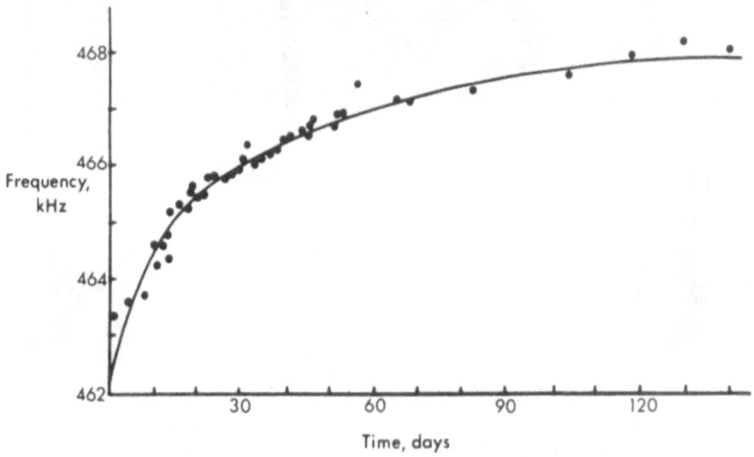


FIGURE 7. Variation of center frequency of telemetry capsule following implant.

niques or indwelling transducers are useful for monitoring: (1) bladder pressure as a means to measure volume, (2) bladder pressure as a means to determine efficacy of stimulation techniques, and (3) function of an implanted artificial sphincter system.

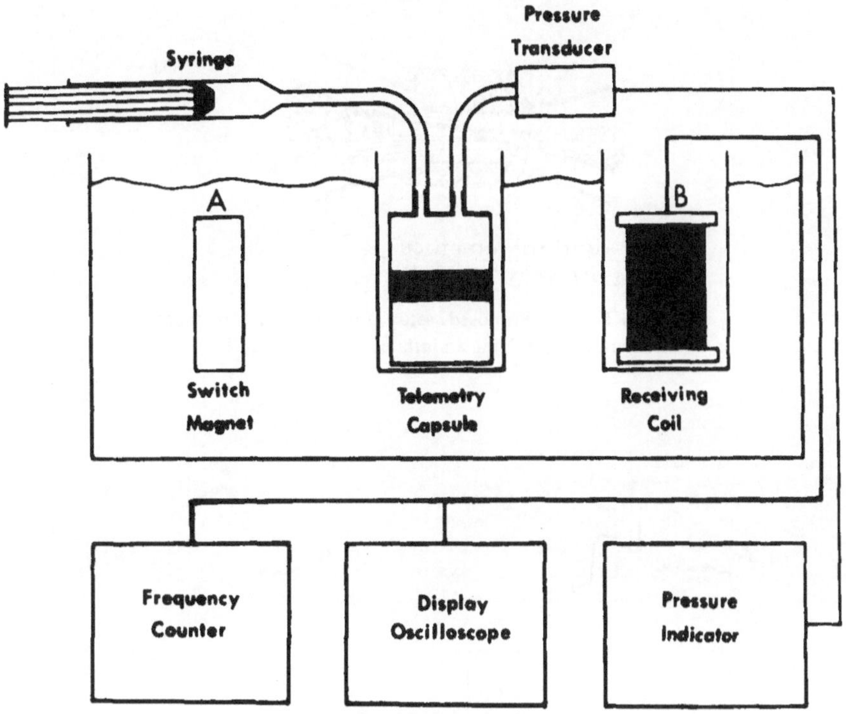


FIGURE 8. Experimental setup to regulate environment of capsule.

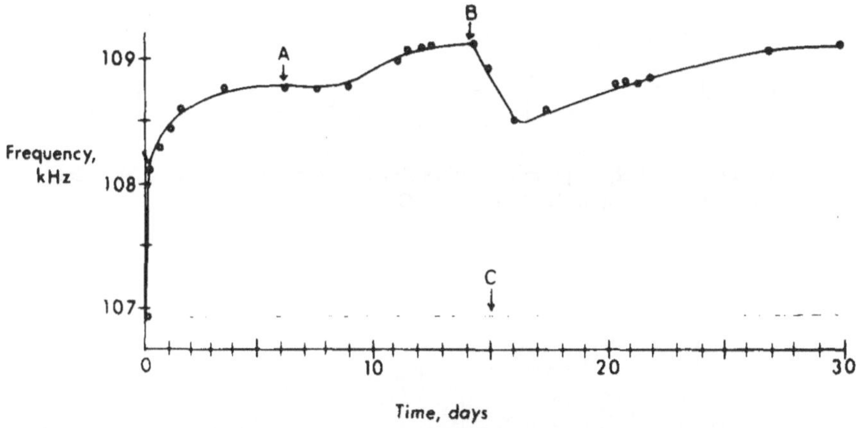


FIGURE 9. Center frequency shift after temperature increase (A), after filling capsule with 25% Hypaque® (B), and after immersing entire capsule in saline (C).

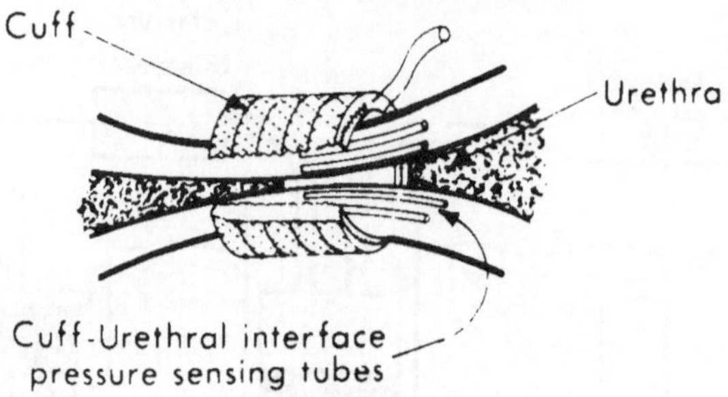


FIGURE 10. Proposed setup for measuring pressure distribution underlying an inflated sphincteric cuff.

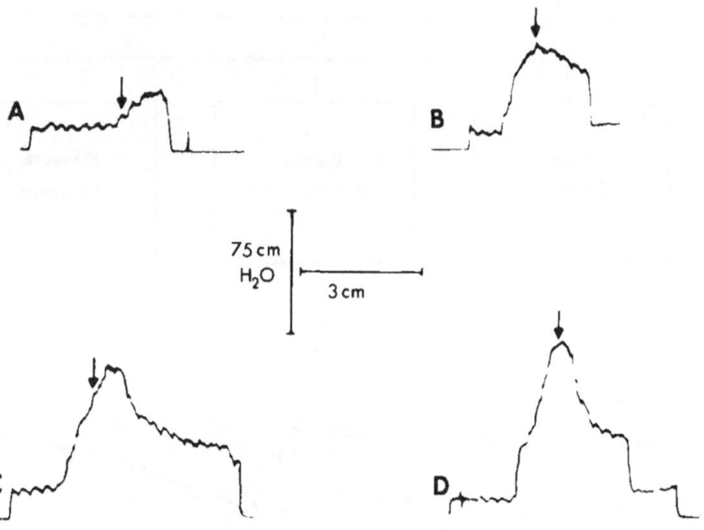


FIGURE 11. Urethral pressure profiles obtained with cuff inflated to: (A) 0 cm $H_2O$ gauge, (B) 50 cm $H_2O$, (C) 85 cm $H_2O$, and (D) 100 cm $H_2O$.

## DIALOGUE

**Dr. Neuman**: What was the technique for getting your pressure profile (Figure 11) under the cuff? How do you use this technique to get a profile of the urethra without the cuff – would you comment on this?

**Dr. Timm**: Yes, well the techniques we used in both cases were the same. The initial technique was developed some years ago in England. A tube with a small port in the side, about 1 ml in diameter, is drawn through the urethra while saline or some fluid is passed through the port at a very slow rate. The pressures required to infuse the fluid at a constant rate are monitored continuously.

So, if you infuse at a constant infusion rate and ignore losses along the tube, you have the tissue pressure at the site of the side port of the catheter.

**Dr. Neuman**: You then slowly withdraw?

**Dr. Timm**: You slowly withdraw at a controlled rate. This has been done both in patients with cuffs and without. In fact, this is one of the ways we use to measure if the device is performing properly.

**Dr. Neuman**: Do you measure intra-vesical pressure, and do you make the pressure profile?

**Dr. Timm**: Sometimes we do and sometimes we don't. It depends on the information we are trying to get. If we are trying to gain information about the cuff, it is not of any particular importance. Trying to measure the function of the bladder and its interrelation with the urethra makes it more critical. Sometimes a third pressure is also measured, namely the intra-abdominal pressure, so that it can be subtracted from the intra-bladder or intra-vesicular pressure.

## REFERENCES


1. **Timm, G. W. and Bradley, W. E.,** Technologic and biologic considerations in neuro-urologic prostheses development, *IEEE Trans. Biomed. Eng.*, BME-20, 208, 1973.
2. **Gleason, D. M. and Lattimer, J. K.,** A miniature radio transmitter which is inserted into the bladder and which records voiding pressures, *J. Urol.*, 87, 507, 1962.
3. **Dreher, R. D., Timm, G. W., and Bradley, W. E.,** Bladder volume sensing by local distension measurement, *IEEE Trans. Biomed. Eng.*, BME-19, 247, 1972.
4. **Timm, G. W. and Bradley, W. E.,** Electrostimulation of the urinary detrusor muscle to effect contraction and evacuation, *Invest. Urol.*, 6, 562, 1969.
5. **Timm, G. W.,** An implantable incontinence device, *J. Biomech.*, 4, 213, 1971.
6. **Scott, F. B., Bradley, W. E., Timm, G. W., and Kothari, D.,** Treatment of incontinence secondary to myelodysplasia by an implantable prosthetic urinary sphincter, *South. Med. J.*, 66, 987, 1973.
7. **Timm, G. W., List, J. S., Bradley, W. E., and Scott, F. B.,** Frequency stability of an implantable pressure telemetry capsule, *IEEE Trans. Biomed. Eng.*, BME-22, 250, 1975.
8. **Bradley, W. E. and Timm, G. W.,** Combined electromyographic and gas urethral pressure profilometry, *J. Urol.*, 115, 433, 1976.
9. **Ghoneim, M. A., Rottembourg, J. L., Fretin, J., and Susset, J. G.,** Urethral pressure profile: Standardization of technique and study of reproducibility, *Urology*, 5, 632, 1975.
10. **Brown, M. and Wickham, J. E. A.,** The urethral pressure profile, *Br. J. Urol.*, 41, 211, 1969.
11. **Delaney, J., Broadie, T., Timm, G., and Bradley, W.,** A prosthetic sphincter for the gastrointestinal tract, *J. Surg. Res.*, 16, 204, 1974.
12. **Kothari, D. R., Timm, G. W., Frohrib, D. A., Bradley, W. E., and Scott, F. B.,** An implantable fluid transfer system for treatment of impotence, *J. Biomech.*, 5, 567, 1972.
13. **Scott, F. B., Bradley, W. E., and Timm, G. W.,** Management of erectile impotence: Use of an implantable inflatable prosthesis, *Urology*, 2, 80, 1973.

# *Manufacturer's Considerations of Indwelling Pressure Transducers*

# PUBLISHED SPECIFICATIONS OF COMMERCIAL PRESSURE TRANSDUCERS

**Scott Boettcher**

The number of pressure transducers approved for medical applications is extremely limited. Among this group, an even smaller number are suitable for implantation and are used for direct measurement of biological pressures. The primary restriction for implantable transducers is physical size, because many commercial sensors are available with sensitivity that meets the requirements for implantation. Few, however, are capable of being packaged for internal applications.

The list that follows has been divided into three categories. The first group includes sensors designed for internal use or implantation. The second category contains transducers intended for biomedical applications, but these are mounted externally and are connected to the system via catheters or conductive mediums. The last group has been included for comparison purposes and represents a sample of the smallest sensors commercially available. These transducers are not packaged or specifically intended for medical use, but they are approximately the same size as those used for implantation.

The list contains the published specifications of sensors available as of January 1976. Wherever possible, conversions have been made to common units of measure. In all cases the best condition specifications have been used. Some sensors, however, demonstrate a sensitivity shift with temperature, and this characteristic may not be included in the manufacturer's specifications. The list is representative rather than comprehensive and should not be considered a definitive survey of all sensors currently available.

**Commercial Pressure Transducer Comparison**

| Manufacturer model no. | Size | Sensitivity | Compensated temperature coefficient | Linearity and hysteresis | Stability |
|---|---|---|---|---|---|
| | | | *Internal or implantable type* | | |
| Konigsberg | | | | | |
| P 3.5 | 3.5 mm diam | 5 μV/mmHg-V | 0.35% reading/°C 32–42°C | ±0.6% FS 0–300 mmHg | no specifications |
| P 4.0 | 4.0 mm diam | 6.6 μV/mmHg-V | 0.35% reading/°C 32–42°C | ±0.6% FS 0–300 mmHg | 10–30 mmHg/month |
| P 4.5 | 4.5 mm diam | 8.3 μV/mmHg-V | 0.35% reading/°C 32–42°C | ±0.6% FS 0–300 mmHg | 10–30 mmHg/month |
| Kulite® | | | | | |
| MCP | 1.7 mm diam (5 French) | 10 μV/mmHg–V | 0.18 mmHg/°C | ±1% FS or ±3 mmHg 0–300 mmHg | 3–4 mmHg/hr |
| MPSL | 10.2 mm diam | 10 μV/mmHg-V | 0.028 mmHg/°C | ±0.5% FS 0–300 mmHg | repeatability 0.5 mmHg |
| Millar | | | | | |
| Micro-Tip® | 1.7 mm diam (5 French) | 24 μV/mmHg-V | ±0.1 mmHg/°C | ±0.5% FS –300 + 400 mmHg | Better than 1 mmHg/hr |
| Honeywell® | | | | | |
| 3110 (Esophageal only) | 5 mm diam | 5 μV/mmHg-V | 1 mmHg/°C | ±1% FS 0–300 mmHg | Better than 20 mmHg/8 hr |
| | | | External types | | |
| Bell & Howell® | | | | | |
| 4–327 I | 25 mm diam | 5 μV/mmHg-V | less than 0.23 mmHg/°C | ±0.5% FS 0–300 mmHg | no specifications |
| Bentley® Trantec | | | | | |
| Arterial | 25 mm diam 30 mm long | 5 μV/mmHg-V | no specifications | 1% FS 0–300 mmHg | no specifications |
| Venous | 25 mm diam 30 mm long | 25 μV/mmHg-V | no specifications | 1% FS –10 + 60 mmHg | no specifications |
| Statham® | | | | | |
| VBPS (Blood pressure monitor) | 35 mm diam 35 mm long | 40 μV/mmHg-V | ±0.054 mmHg/°C | ±1% FS 0–300 mmHg | no specifications |

**Commercial Pressure Transducer Comparison (continued)**

| Manufacturer model no. | Size | Sensitivity | Compensated temperature coefficient | Linearity and hysteresis | Stability |
|---|---|---|---|---|---|
| | | | Nonmedical sensors | | |
| Entran | | | | | |
| EPI-080-5 | 2 mm diam 6.4 mm long | 11 $\mu$V/mmHg-V | ±0.3 mmHg/°C | ±0.5% FS 0–260 mmHg | repeatability 0.3% FS |
| Type H-Foil | 3.2 mm diam 5 mm long | 0.77 $\mu$V/mmHg-V | 0.45 mmHg/°C | 1% FS 0–500 mmHg | repeatability 0.25% FS |
| Type H-Semi | 3.2 mm diam 5 mm long | 9.7 $\mu$V/mmHg-V | 0.45 mmHg/°C | 1% FS 0–500 mmHg | repeatability 0.25% FS |
| Type H | also available as wafer 0.86 × 3.8 × 9.5 mm | | | | |
| Setra Systems | | | | | |
| 240TC 1/8 in. | 3.2 mm diam | 0.6 mV/mmHg-V (with amp) | 0.023 mmHg/°C (sensor only) | ±0.5% FS 0–260 mmHg | no specifications |
| 240TC 1/4 in. | 6.4 mm diam | 17 mV/mmHg-V (with amp) | $0.5 \times 10^{-3}$ mmHg/°C (sensor only) | ±0.5% FS 0–5.2 mmHg | no specifications |

# MEDICAL PRESSURE DEVICES AND SYSTEMS

**Jess W. Burns**

Applications of external pressure transducers in indwelling systems are reviewed. One proposed transducer concept for long-term implantation in the cardiovascular system is described.

**Mr. Burns**: My participation in medical instruments technology has been almost completely with external transducers, but I will present a couple of applications here relating to indwelling service. First, I must state that many of the good ideas in Statham medical instruments have come from physicians. If this head (transducer pressure dome) looks familiar, I think you have seen it copied. This is about the sixth or seventh in the series which started back about 1950 and was suggested by Dr. Wood. He said saline should be introduced in one (side) port and flushed out the other (center) port, that is, to fill the pressure dome. I have to chide him a little because you should not flush through it later on (after filling and debubbling). That is bad practice.

**Dr. Wood**: Well, that is because your gauge is so sensitive.

**Mr. Burns**: Well, there is a way around that.

**Dr. Wood**: You should make a by-pass through that gauge.

**Mr. Burns**: I do not think you have tried the very latest model. The P23ID is pretty good.

Regarding applications, I have here an accessory called the Richmond (subarachnoid) screw. You may be familiar with it. It comes with handle and twist drill. Statham does not sell it, but does have a transducer that is commonly used with it. It requires a ¼-in. drill, which is used to drill a hole in the skull. The dura is exposed and cut, and the screw is threaded in and connected to a stopcock. The system is filled with saline and connected to a Statham® P50A transducer. At this point the system is zeroed in the same manner that Dr. Wood suggested. Usually the zero reference is established at the location of the transducer, but in this instance it is actually at the bottom of the screw.

This same transducer is in another set-up, with a similar hole in the skull. The transducer is inserted in this fixture (a metal tube-like body, with the inside shaped to receive a P50A transducer with its diaphragm flush with the end that has threads to enter the hole in the skull) and threaded in. The dura, itself, becomes the sensing diaphragm. You do not go through it. In effect, that is a force-measuring system since no liquid is required, but it can be used to sense intracranial pressures.

One brief comment on the state-of-the-art: it seems to be an accepted fact that today everyone is going to solid state pressure transducers, notably the silicon type, but there is another technology we probably will be using. That is, thin-film technology, which provides extremely stable elements and can be made a lot less temperature sensitive than silicon-based transducers.

Another application was developed by Jack Chambers at Statham, who worked on the left ventricular by-pass and is still working on certain phases of the artificial heart project. He was using a Dacron® velour and found that one could, in a controlled manner, grow a blood clot across the surface of this material. It was impermeable to blood, but would pass water and sodium and chloride ions. He started a program that, unfortunately, was not followed on a long-term basis. He set up a small, liquid-filled test chamber with a P23 transducer connected to it. The velour covered an opening in the chamber. Blood was pumped through a channel on the other side of the velour.

He was able to show that water and ions would migrate through the velour and that the fabric could develop a controlled fibrin deposit. In effect, there is a bubble-free liquid between the pressure transducer and the velour. The velour may have a fairly large cross-sectional area which would result in a minimal pressure drop through it. As the pressure transducer may be extremely stiff, the isolation barrier (fibrin-covered velour) hardly needs to move. The pressure transducer can therefore maintain its initial characteristics for an unlimited time (eliminating the usual problem of fibrin overgrowth changing the sensitivity of an exposed sensing diaphragm).

The project continued through the investigation of the characteristics of fibrin-coated velour and other membrane materials. A test fixture was fashioned so that four membrane materials could be exposed to flowing blood at one time. The proposed transducer incorporating a controlled endothelial growth is shown in Figure 1. It offers a promising approach for long-term implantation of a functional pressure transducer.

The question came up earlier: "How do you get the little chips out of the wafers?" Without giving away any trade secrets of Kulite, in my brief brush with chip production, there were three common ways: you can saw then with a diamond saw, etch them, or nick them with a diamond and break them, which gives you rough edges. There are probably three or four other ways to do it.

**Mr. Konigsberg**: If you drop them on the floor, that is called the stochastic method.

**Mr. Burns**: This has been rather brief, but I think it fairly well covers the applications I had in mind.

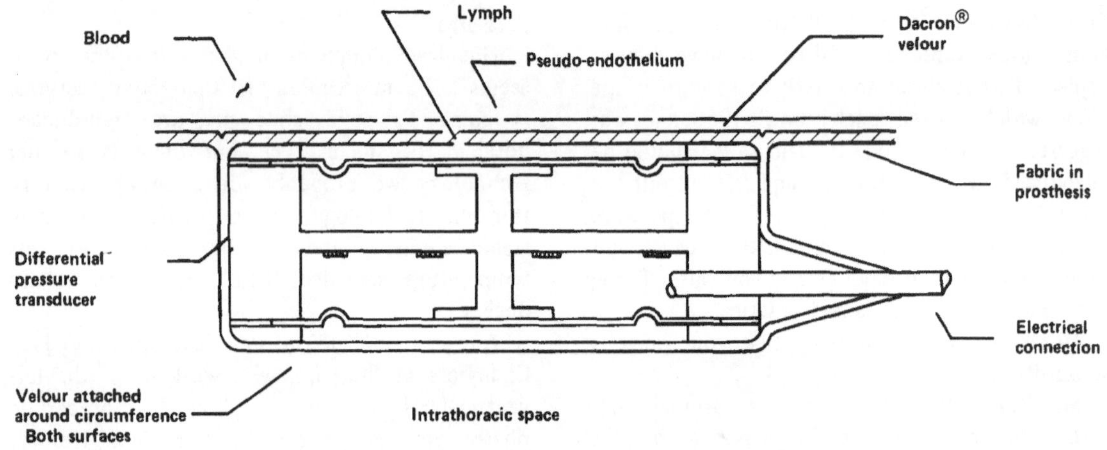


FIGURE 1. Pressure transducer design allowing fibrin attachment.

*Research and Development of Indwelling Pressure Transducers*

# PRINCIPLES OF PRESSURE TRANSDUCERS

**Wen H. Ko**

Physical principles that can be used for pressure transducers are reviewed, including thermal, electrostatic, electrokinetic, gas discharge, and mechanical deformation effects of pressures. The principles used for biomedical pressure transducers are summarized, including piezoelectric, piezoresistive, semiconductor junction, and mechanical deformation. Current problems in designing biomedical pressure transducers are discussed briefly.

A short review of the principles of pressure transducers, particularly those used in biomedical measurements, should be included at this point. In a previous chapter, Dr. L. Geddes states that all the medical pressure transducers are, in one way or another, devices which translate mechanical deformation into electrical signals. This was true in the past and is still true today considering the broad interpretation of deformation. The mechanical deformation can be visible in the classical sense or in microdeformation, as in the crystal structure. Pressure, however, has many effects besides mechanical deformation, and some of these phenomena will be discussed as follows. These other effects have been used to measure pressure for industrial applications or scientific research, but generally have not been used for biomedical purposes since they are not in a useable pressure range or sensitivity.

**Thermal effect** – Certain materials will change their boiling temperature or melting temperature according to pressure. Since we can measure temperature accurately, we can secondarily measure pressure.

**Electrostatic effect** – Some materials change dielectric constant with the pressure; therefore, one can build a pressure-sensitive capacitor which does not require movement of the electrodes to measure the pressure. As shown in Figure 1, a capacitor could be constructed with this type of dielectric material. Varying the pressure on this material will alter its dielectric constant, thereby changing the total capacitance of the device.

**Electrokinetic effect** – This effect is similar to the streaming potential many of us have observed associated with fluid flow. When the fluid flows through a small tube, there is an exchange of electric charge between the solid material and the

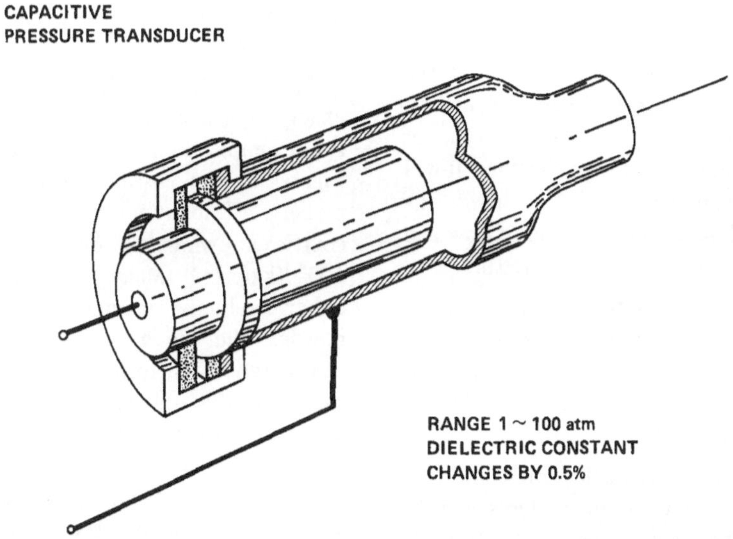


FIGURE 1.

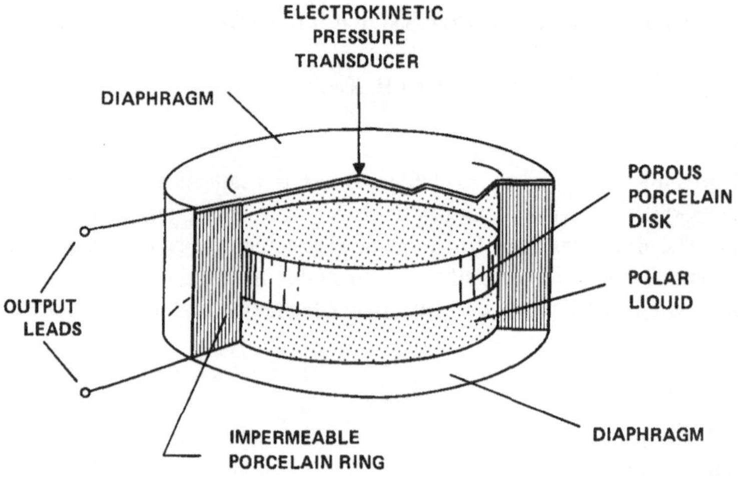


FIGURE 2.

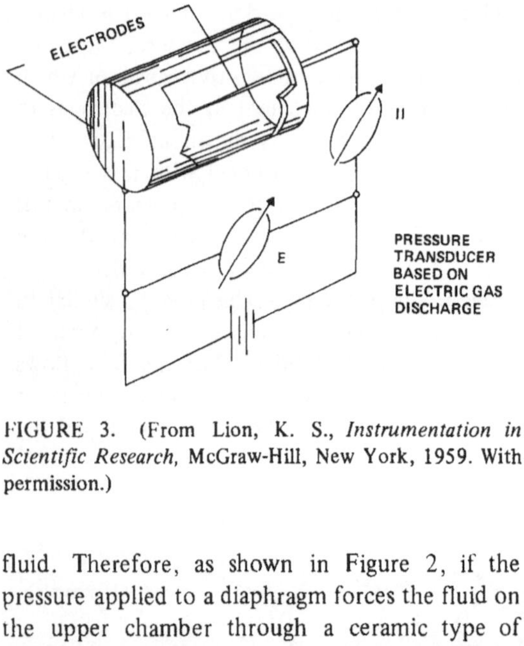


FIGURE 3. (From Lion, K. S., *Instrumentation in Scientific Research,* McGraw-Hill, New York, 1959. With permission.)

fluid. Therefore, as shown in Figure 2, if the pressure applied to a diaphragm forces the fluid on the upper chamber through a ceramic type of porous wafer and goes to a lower chamber, it will produce a positive or negative voltage across the electrodes, depending on the direction of flow and its magnitude which is related to the pressure.

**Gas discharge voltage effect** – The discharge voltage of a gas discharge tube depends on the gas pressure, as shown in Figure 3. By measuring the voltage at a given current, the pressure inside the chamber can be measured.

The range of measurable pressure, the sensitivity of the above-mentioned effects, and the convenience of the devices have not yet made them useable for medical transducer applications. Currently used biomedical pressure transducers operate on the following major principles:

**Piezoelectric effect** – When pressure is applied to certain ceramic or crystal materials (either compression or tension), electric charges are generated on both surfaces due to polarization of the molecules of the material. As shown in Figure 4, the equivalent circuit of the device is a voltage generator proportional to the pressure, in a series with a capacitor and a resistor. This type of transducer can be used to measure alternating pressure, but not steady pressure. Figure 5 shows an industrial piezoelectric pressure transducer. When under pressure, the transducer element generates charges and a voltage across the terminals. For an AC type of signal, this device is very sensitive. It can generate a high voltage with a high output impedance.

**Piezoresistive transducer** – Many materials will change resistance in response to pressure changes. For example, in a simple structure as shown in Figure 6, the resistor changes its value with pressure. Rubidium has a resistance pressure coefficient which is quite large, about $1 \times 10^{-4}/(\mathrm{kg/cm^2})$. Many polymeric materials are either conductive or can be mixed with conductive powder, and can be molded into rods or other configurations to form resistors. As illustrated in Figure 7, the rod changes dimensions and thus changes resistance. The carbon pile or carbon powder has been used to measure acoustic pressure by its pressure sensitivity of contact resistance. This principle has long been used in microphones.

**Silicon strain gauges** – These devices vary their resistance with strain or the pressure that causes

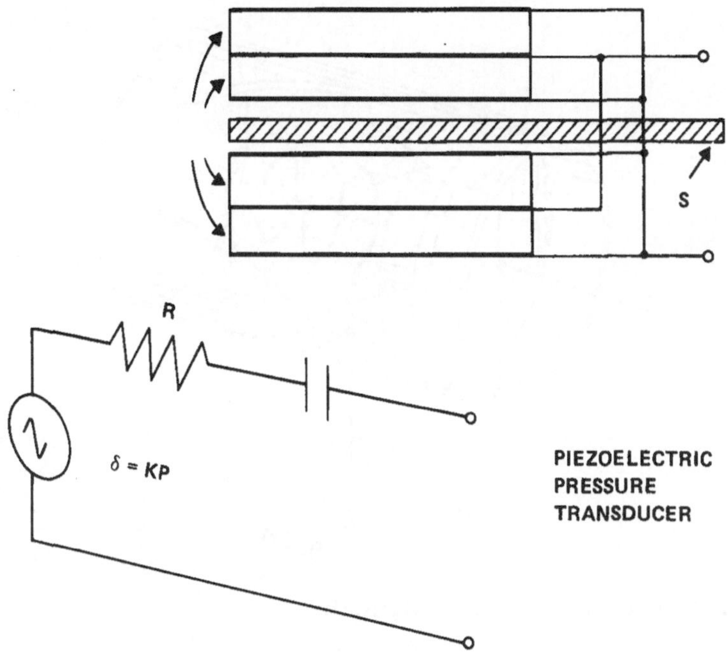


FIGURE 4. (From Lion, K. S., *Instrumentation in Scientific Research*, McGraw-Hill, New York, 1959. With permission.)

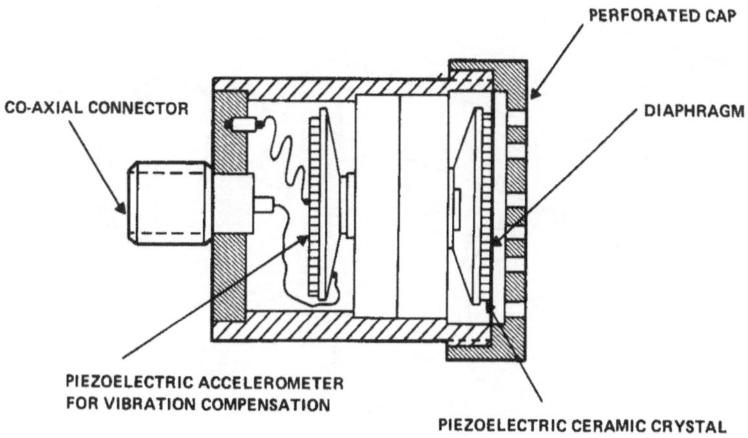


FIGURE 5. (From Norton, H. N., *Handbook of Transducers for Electronic Measuring Systems*, Prentice-Hall, Englewood Cliffs, New Jersey, 1969. With permission.)

the strain. Usually when the bridge formed by four silicon resistors is excited with external DC or AC sources, the bridge output will be proportional to strain or pressure. There are major problems associated with these devices including zero drift, thermal drift, sensitivity change with temperature, packaging, and so on. A more detailed discussion on silicon strain gauges is provided in other chapters in the section on "Research and Development in Indwelling Pressure Transducers."

**Semiconductive devices** – In addition to silicon resistors, pn junction devices are also pressure sensitive. One of the available devices is the pnp transistor. When the base-emitter junction is stressed, the current gain changes. This principle is used in the commercial product made by Pitran, whose structure is illustrated in Figure 8. Basically, the device contains a metal diaphragm coupled to a stylus resting on the base-emitter junction of the transistor. This arrangement acts as a mechanical amplifier to concentrate the force on the junction. Changes in force alter the current gain and the

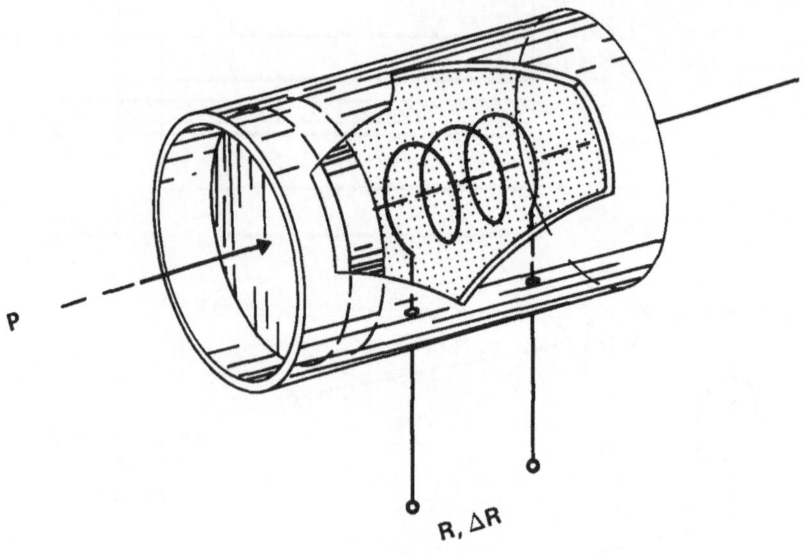


**RESISTIVE PRESSURE TRANSDUCER**

FIGURE 6.

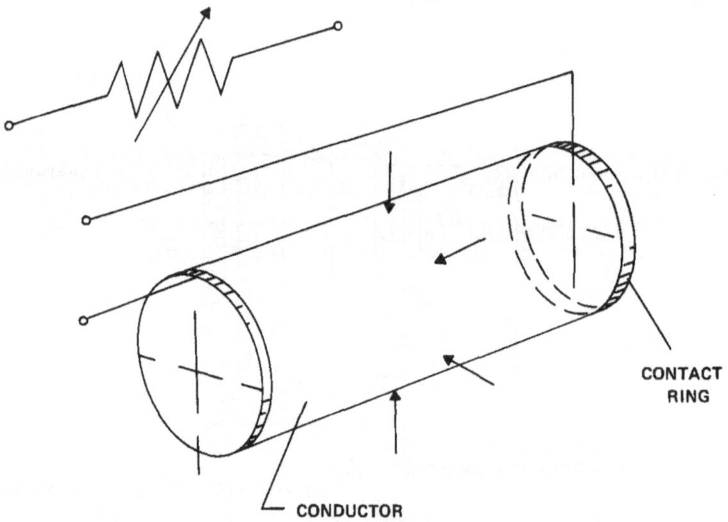


FIGURE 7.

collection current of the transistor. Its applications are limited by its sensitivity to mechanical damage and fatigue of the device. Another device principle which has been explored by NASA and the Engineering Design Center at Case Western Reserve University is the pressure sensitivity of the tunnel diode. In the tunnel diode, the current voltage characteristics, as shown in Figure 9, have an N-shaped wave form. The peak current, $I_p$, is quite sensitive to pressure. For testing purposes, such a device is mounted on the cantilevered beam. When a force representing pressure is applied to the beam, it causes the peak current to vary. The percentage of change of the peak current versus tension or compression strength is quite linear, as shown in Figure 10. The tunnel diode has the advantage of being a low voltage and low impedance device, so that the problem of protecting the device from body fluid can be greatly reduced. However, no extensive study has been completed and there is no available commercial product at the present time.

PITRAN PRESSURE TRANSDUCER

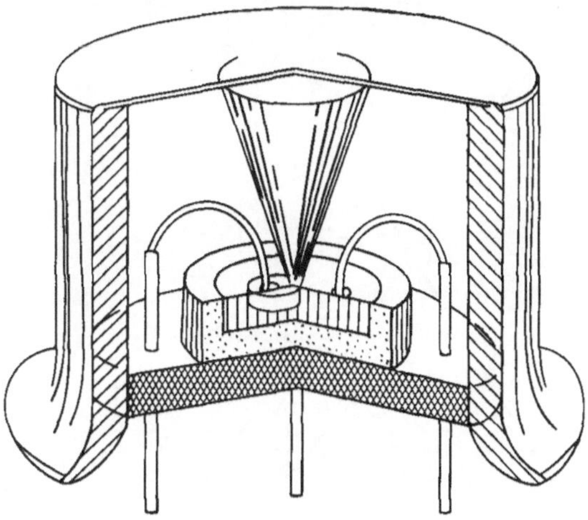

FIGURE 8.

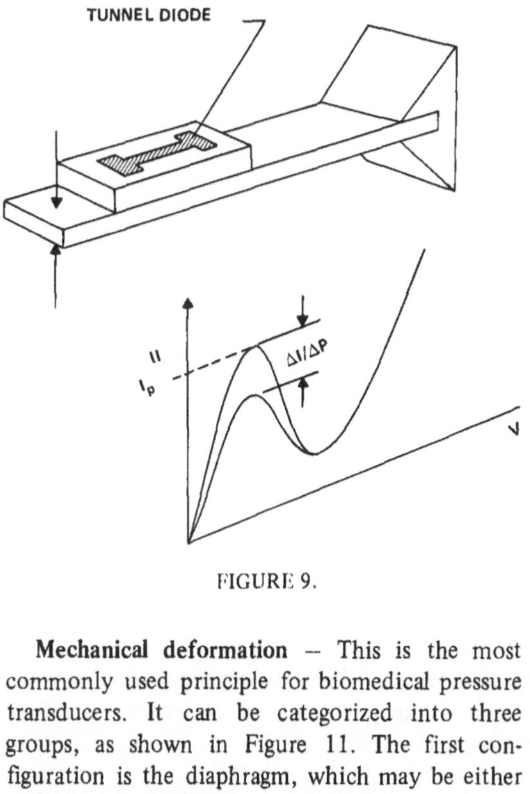


FIGURE 9.

**Mechanical deformation** – This is the most commonly used principle for biomedical pressure transducers. It can be categorized into three groups, as shown in Figure 11. The first configuration is the diaphragm, which may be either flat or corrugated. A second geometric form is the bellow, which is really a folded diaphragm used to increase the sensitivity and reduce the area or volume. A mechanical bellow changes its longitudinal dimensions when the pressure changes. The third structure is the tube which can be either twisted or curved. The pressure change causes the tube to change curvature or to rotate.

Most of the mechanically deformed biomedical pressure transducers are used to convert the displacement into electrical signals that can be processed later. The common principles are summarized as follows:

**Capacitive** – Figure 12 shows a device where the center diaphragm has two capacitances related to the upper fixed plate and the lower fixed plate. When the differential pressure is applied, the diaphragm deflects; therefore, these two capacitances and the bridge circuit become unbalanced resulting in a voltage output.

**Inductive** – Figure 13 indicates a scheme to convert the mechanical deformation caused by pressure changes into the up and down movement of a ferrite rod inside a differential transformer to generate an unbalanced differential voltage. Many other mechanical arrangements can be made so that mechanical deformation can be translated into variable inductance.

**Resistive** – The mechanical movement can also be coupled to a potentiometer to produce a variable resistance device.

**Optical** – The deformation of a diaphragm can be measured by an optical means, for instance, by bouncing a light against the diaphragm and picking it up by a detector, as shown in Figure 14. When the diaphragm deflects, the amount of light that can be picked up by this detector will vary; therefore, one can have a noncontact type of measurement of deformation. The main problem is how to make this diaphragm stable and reliable.

**Vibrational** – When the tension of a wire is proportional to pressure, and the wire is suspended in a magnetic field, the vibration frequency of the wire will vary with the tension or the pressure.

**Volumetric** – As shown in Figure 15, when the diaphragm is deflected, it causes the fluid in a filled chamber to move up or down from a fluid column. Various methods of detecting the height of the fluid column by optical, inductive, or other means can be used to translate the column height and pressure into electrical signals.

**Fluid column manometer** – The most fundamental means of measuring pressure is by the differential height of the fluid columns under differential pressure, as shown in Figure 16. This is usually used as the calibration method.

Principles that have been explored for pressure measurements, particularly those phenomena which have been used in the pressure measurement for biomedical purposes, were briefly outlined above. At the present, the major problems in designing biomedical pressure transducers are as follows:

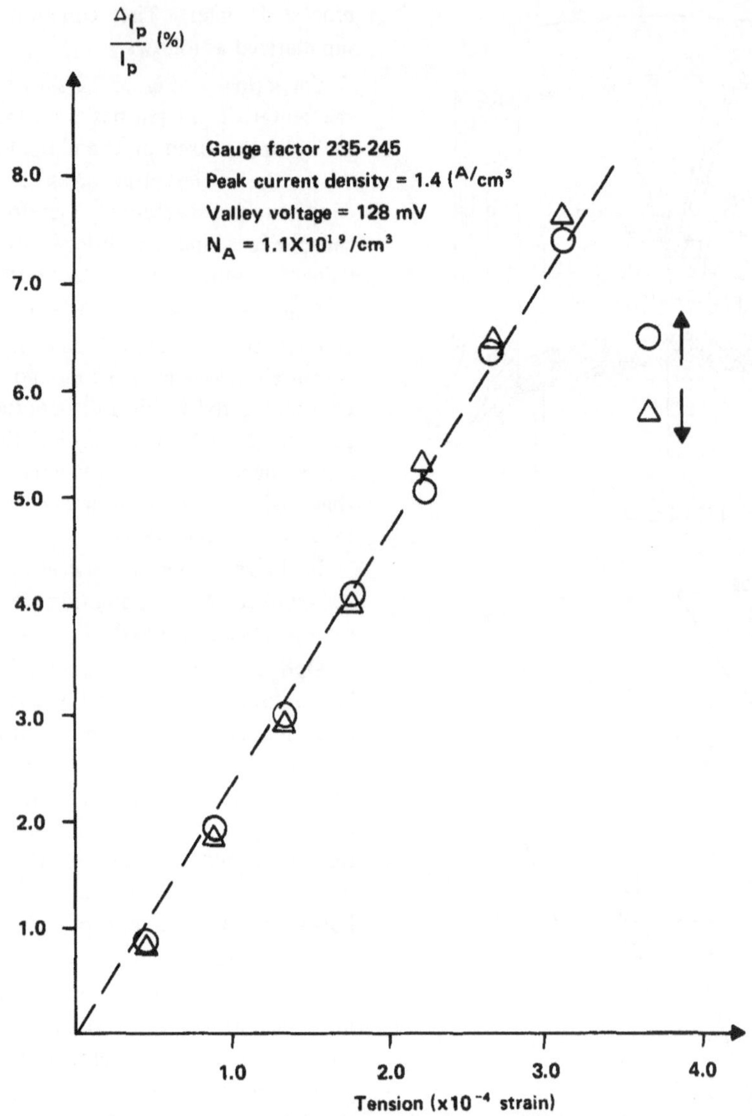

Beam tension strain effect on $I_p$ of SHP-2-A-8B4

FIGURE 10.

1. The size and weight must be small enough so that they do not disturb the experimental subjects. The acceptable value depends on the type of experiments to be performed.

2. For implanted transducers, depending on the time duration of the measurements, the hysteresis and the zero reference stability are important considerations. Since there is no atmospheric pressure available for reference, absolute pressure would have to be measured for the implant devices.

3. The method of supplying power to the indwelling pressure transducer is also a problem. It would be desirable to develop techniques to minimize the power of the transducer to a few microwatts.

4. In designing a complete transducer system, the last but most important problem is how to process the signal, reduce the noise, and increase the stability so that one can measure small pressures, such as venous pressure, with good accuracy over a long time period.

A conversion table of various pressure units is given in Table 1.

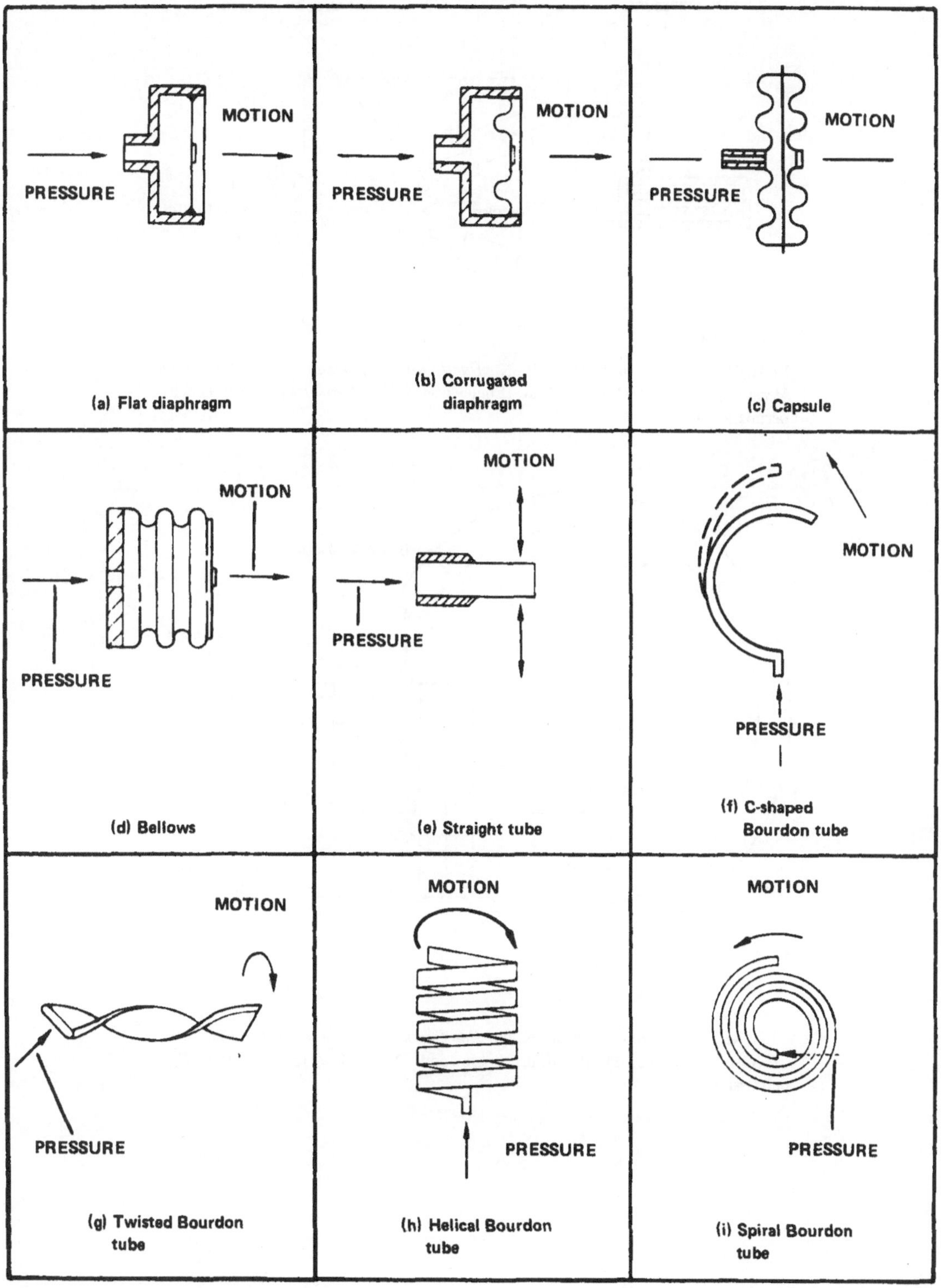


FIGURE 11. (From Norton, H. N., *Handbook of Transducers for Electronic Measuring Systems*, Prentice-Hall, Englewood Cliffs, New Jersey, 1969. With permission.)

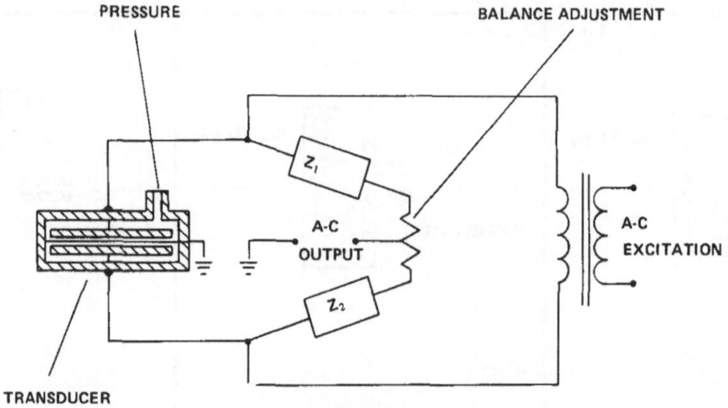


FIGURE 12. (From Norton, H. N., *Handbook of Transducers for Electronic Measuring Systems,* Prentice-Hall, Englewood Cliffs, New Jersey, 1969. With permission.)

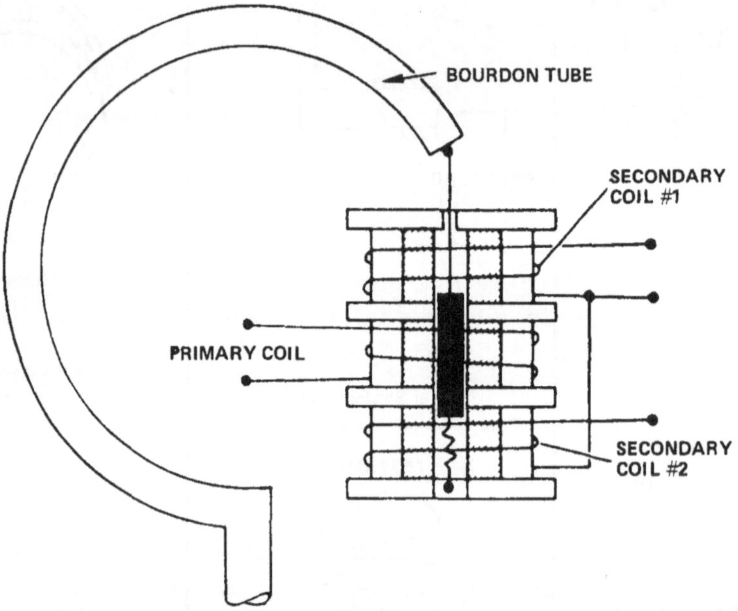


FIGURE 13. (From Norton, H. N., *Handbook of Transducers for Electronic Measuring Systems,* Prentice-Hall, Englewood Cliffs, New Jersey, 1969. With permission.)

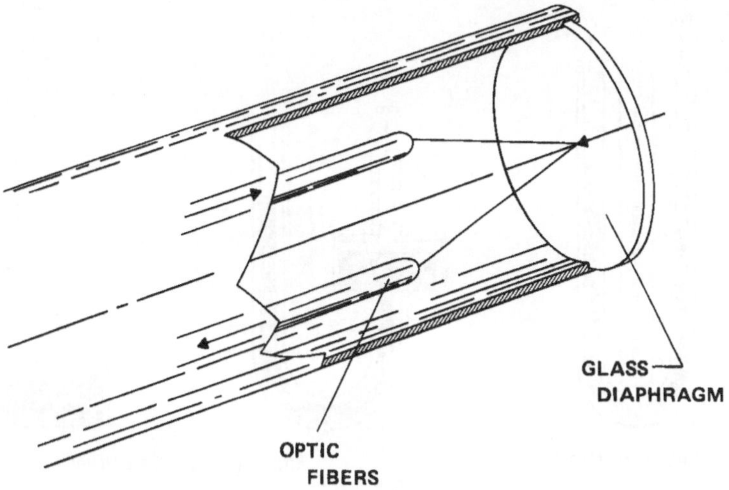


OPTICAL CATHETER TIP PRESSURE TRANSDUCER

FIGURE 14.

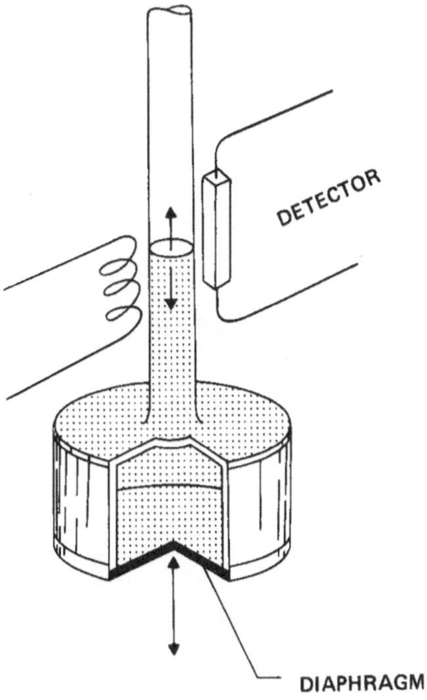


FIGURE 15.

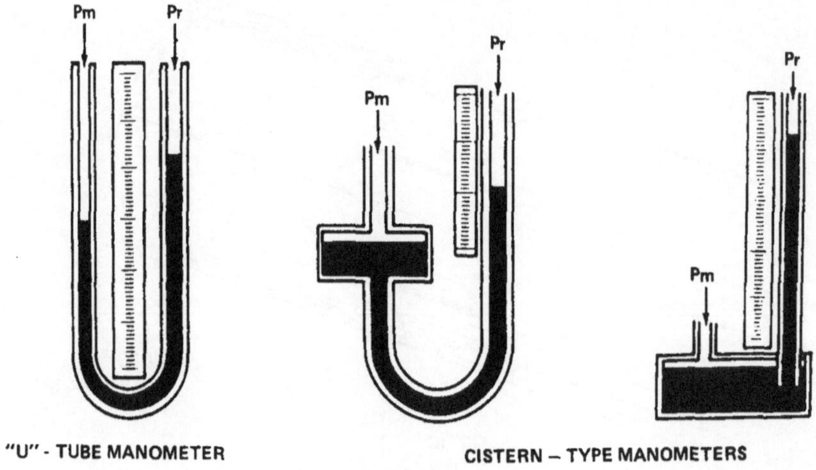


Manometers.

FIGURE 16. (From Norton, H. N., *Handbook of Transducers for Electronic Measuring Systems*, © 1969, p. 481. Reprinted by permission of Prentice-Hall, Inc., Englewood Cliffs, New Jersey.)

TABLE 1

**Units of Pressure Measurement**

| | psi | $kg/cm^2$ | in. $H_2O$ | cm $H_2O$ | in. Hg | mmHg | $dyne/cm^2$ |
|---|---|---|---|---|---|---|---|
| 1 psi = | – | 0.0703 | 27.67 | 70.13 | 2.036 | 51.715 | 68,947 |
| (1) 1 $kg/cm^2$ = | 14.22 | – | 393.7 | 1,000 | 28.96 | 735.56 | 980,665 |
| (2) 1 in. $H_2O$ = | 0.03613 | 0.00254 | – | 2.540 | 0.07355 | 1.868 | 2,491 |
| (2) 1 cm $H_2O$ = | 0.0142 | 0.001 | .3937 | – | 0.02896 | 0.7356 | 980.64 |
| (3) 1 in. Hg = | 0.4912 | 0.0345 | 13.59 | 34.53 | – | 25.40 | 33,864 |
| (3)(4) 1 mmHg = | 0.01934 | 0.001359 | .5352 | 1.35 | 0.03937 | – | 1,333 |
| (5) 1 $dyne/cm^2$ = | $1.45 \times 10^{-5}$ | $1.02 \times 10^{-6}$ | $4.015 \times 10^{-4}$ | $1.02 \times 10^{-3}$ | $2.953 \times 10^{-5}$ | $7.5 \times 10^{-4}$ | – |
| (6)(7) 1 atmosphere = | 14.696 | 1.033 | 406.8 | 1,033 | 29.92 | 760 | 1,013,250 |

*Notes:* (1) 1 $kg/m^2$ = 10,000 $kg/cm^2$

(2) in. $H_2O$ and cm $H_2O$ are referenced to a temperature of 4°C (39.2°F)

(3) in. Hg, mmHg and $\mu$ Hg are referenced to a temperature of 0°C (32°F)

(4) 1 mmHg = 1,000 $\mu$ Hg

(5) 1 $dyne/cm^2$ = 1 microbar = $1 \times 10^{-6}$ bar

(6) 1 bar = 0.987 atmospheres

(7) Atmospheres are more precisely called normal atmospheres

# REFERENCES


1. *American Standard Procedures for Calibration of Electro-Acoustic Transducers, Particularly Those for Use in Water*, American Standards Association, New York, 1958.
2. **Cobbold, R. S.,** *Transducers for Biomedical Measurements: Principles and Applications,* Interscience, New York, 1974.
3. **Harvey, G., Ed.,** *ISA Transducer Compendium,* Instrument Society of America, Pittsburg, 1969.
4. **Herceg, E. E.,** *Handbook of Measurement and Control: An Authoritative Treatise on the Theory and Application of the LVDT,* Schaevite Engineering, Pennsauken, 1972.
5. **Lion, K. S.,** *Instrumentation in Scientific Research: Electrical Input Transducers,* McGraw-Hill, New York, 1959.
6. **Neubert, H. K.,** *Instrument Transducers: An Introduction to Their Performance and Design,* Clarendon Press, Oxford, England, 1963.
7. **Norton, H. N.,** *Handbook of Transducers for Electronic Measuring Systems,* Prentice-Hall, Englewood Cliffs, New Jersey, 1969.
8. **Slemon, G. R.,** *Magnetoelectric Devices: Transducers, Transformers and Machines,* Interscience, New York, 1966.
9. **Doebelin, E. O.,** *Measurement Systems: Applications and Design,* McGraw-Hill, New York, 1966.

# AN IC ABSOLUTE PRESSURE TRANSDUCER WITH BUILT-IN REFERENCE CHAMBER*

**Timothy A. Nunn and James B. Angell**

A new technique for batch fabricating miniature piezoresistive absolute pressure transducers is described. The process makes use of the advanced state-of-the-art of monolithic integrated circuit (IC) processing and anodic bonding, which is a means for hermetically sealing glass to silicon. Advantages of the process include low cost, small size, and improved performance, particularly in regard to long-term drift, which has been measured at <1 mmHg/month.

## INTRODUCTION

In preceding sessions of the workshop, several suggestions were made for improving the stability and decreasing the cost of indwelling pressure tranducers. Batch fabrication was proposed as a means for reducing the cost, and the use of inherently stable materials for constructing the pressure transducer was suggested for improving the stability. Both of these techniques were used to build the miniature absolute pressure transducer (1.0 × 1.3 × 0.4 mm) developed at Stanford University. The transducer is batch fabricated using standard integrated circuit (IC) processing techniques, along with anodic bonding.[1] The anodic bonding is a means for hermetically sealing Pyrex® glass to silicon.

## DESCRIPTION OF TRANSDUCER

The Stanford IC absolute pressure transducer[2] (Figures 1 and 2) consists of a glass cap which is hermetically sealed to a silicon chip. The silicon chip is composed of a thin silicon diaphragm (typically 10 μm thick) which is surrounded by a thick rim of silicon. This thick rim is used for support and protection of the diaphragm. Etched into the top surface of the silicon chip is a well which is 10 μm deep. This well, together with the glass cap, forms the reference pressure chamber. Four p-type resistors are diffused into the top surface of the silicon diaphragm. These resistors are connected by low-resistivity $p^+$ diffusions to form a Wheatstone bridge. The $p^+$ diffusions are also used to make ohmic contacts to the aluminum bonding pads which are evaporated onto the glass cap.

When pressure is applied to the transducer, the diaphragm deflects, causing the resistors to change in value. The change in resistance is linear with pressure, thereby causing the output voltage from the bridge to change linearly with pressure. The four resistors are oriented on the diaphragm such

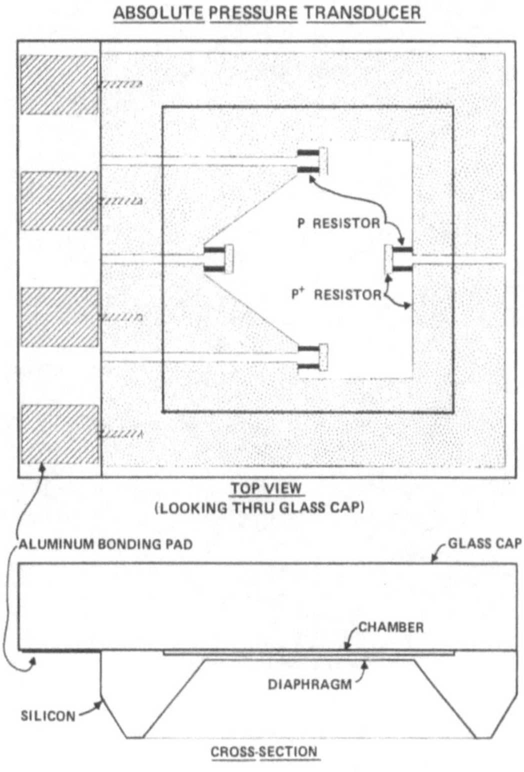


FIGURE 1.

*Paper presented at National Institute of Health sponsored workshop on "Indwelling Pressure Transducers and Systems," Case Western Reserve University, Cleveland, Ohio, December 4 and 5, 1975.

**The support for this work provided by the National Institute of Health via contract #N01-HD3-2774.

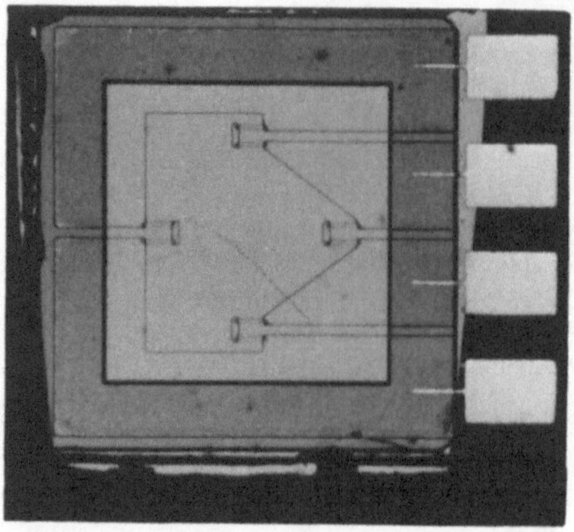

**FIGURE 2A. Absolute pressure transducer (looking through glass cap).**

FIGURE 2B. Absolute pressure transducer on penny for size comparison.

that two opposing resistors increase in value with pressure while the other two opposing resistors decrease an equal amount. Thus the driving point impedance $R_{AD}$ does not change with pressure (Figure 3). However, the p-type resistors have a positive temperature coefficient of resistance, causing $R_{AD}$ to increase with temperature. This change can be used to measure temperature simultaneously with pressure and/or to provide a temperature-sensitive signal to compensate the pressure output, making it insensitive to temperature changes.

Pressure transducer specifications are listed below:

| | |
|---|---|
| Sensitivity | 30 μV/mmHg/Vsupply |
| Range | 0 to 300 mmHg |
| Nonlinearity and hysteresis | less than 0.2% full scale |
| Zero pressure shift | -4 mmHg/°C (uncompensated)<br>-0.1 mmHg/°C (compensated) |
| Sensitivity shift | -0.15%/°C |
| Long-term drift* | < 1 mmHg/mo |
| Calculated natural frequency | > 50 kHz |

## BATCH FABRICATION PROCESS

**1. Forming alignment marks for back-side processing** – The p-type resistors on the top surface of the silicon wafer must be precisely centered on the diaphragm which is formed from the back side of the wafer. To accomplish this, alignment marks are anisotropically etched through the wafer and all masks are aligned with respect to these marks.

**2. Forming the well in the silicon** – The well is anisotropically etched in a $KOH\text{-}H_2O$ solution which has an etch rate of 3 μm/min. Silicon dioxide is used as an etch mask for this process.

**3. Forming the p-type resistors and p$^+$ connectors** – The p-type resistors are diffused using a standard base diffusion which yields a surface resistivity of 100 Ω per square. The p$^+$ diffusion is used as a conductor in this application and is therefore processed to yield a low resistivity (Figure 4).

**4. Forming the back-side etch mask** – The back-side etch mask, which determines the size

* This number is based on results obtained from two pressure transducers tested for 1 month. Additional testing of long-term drift is continuing.

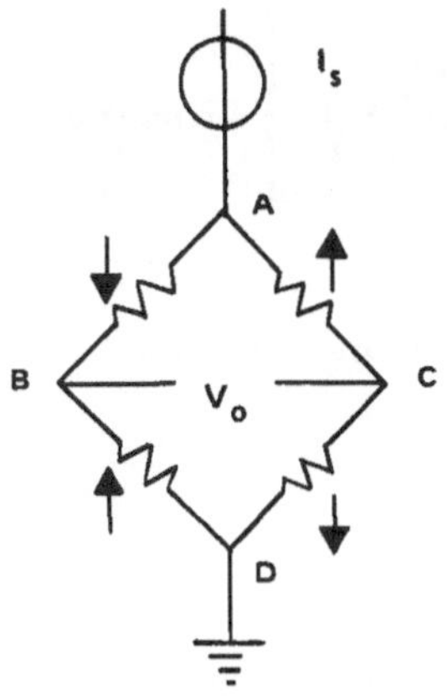


FIGURE 3. Schematic of Wheatstone bridge.

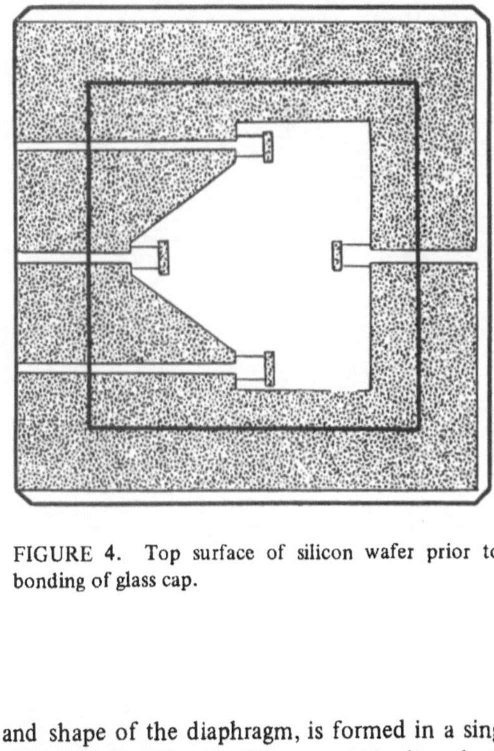

FIGURE 4. Top surface of silicon wafer prior to bonding of glass cap.

and shape of the diaphragm, is formed in a single photolithographic step. The pattern is aligned with the front side pattern by means of the alignment marks formed in step 1.

5. **Forming the bonding pads on the glass** – The bonding pads are formed by evaporating a 0.5-$\mu$m layer of aluminum over the glass plate. A photolithography step protects the aluminum where the bonding pads are desired, and the excess aluminum is then etched away.

6. **Hermetic sealing of the glass to the Si** – The glass plate is hermetically sealed to the silicon wafer by an anodic bond. The glass plate is positioned on the silicon wafer so that the bonding pads on the glass are aligned with the corresponding $p^+$ regions in the silicon. The assembly is then heated on a hot plate to 400°C and a potential of 600 V is applied with the silicon positive with respect to the glass. No additional force is required, since the electrostatic force pulls the glass and silicon into intimate contact and a fusion-type hermetic bond is formed. Simultaneously, the 400°C temperature causes the aluminum pads on the glass to alloy with the $p^+$ regions in the silicon, forming ohmic contacts.

7. **Forming diaphragm** – The diaphragms are now formed by etching away the unprotected silicon from the back side of the wafer. The anisotropic KOH-$H_2O$ etch is also used for this step.

This etching process is carefully monitored, both by timing the etch (etch rate = 3 $\mu$m/min) and by periodically removing the assembly from the etch and measuring the actual diaphragm thickness. Using this method, diaphragms can be reliably etched to within ±1 $\mu$m of the desired thickness.

8. **Separation of the individual transducers** – The glass plate is scribed in the desired shape and the silicon-glass assembly is broken apart into individual absolute pressure transducers. The silicon breaks along V-grooves which are etched into the back side of the wafer during the diaphragm etch.

9. **Lead attachment** – Four leads of 40-gauge multistranded copper wire insulated with polyvinylchloride are attached to the aluminum pads by means of a conductive silver epoxy. The bonds are then covered with nonconductive epoxy for added strength.

## BIOMEDICAL EVALUATIONS

The absolute pressure transducer is now in the prototype testing phase, both in vivo and in vitro. Figure 5 is a pressure transducer trace obtained from an acute in vivo implant in the rabbit oviduct. Further chronic in vivo studies will soon be carried out to measure both intracranial pressure and blood pressure in dogs.

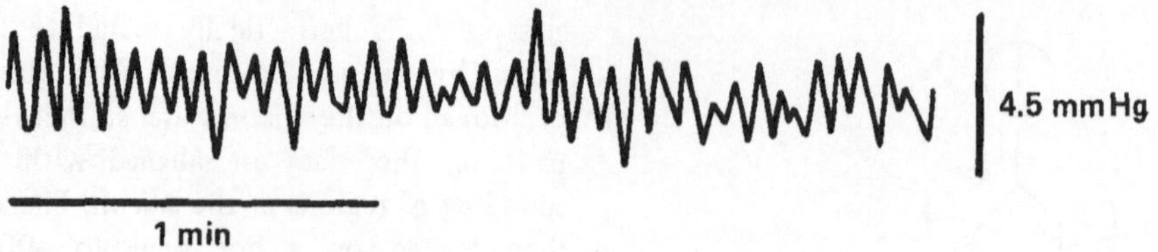


FIGURE 5. Recording made with pressure transducer in rabbit oviduct *in situ*.

# REFERENCES


1. **Wallis, G. and Pomerantz, D. I.,** Field assisted glass-metal sealing, *J. Appl. Phys.*, 40, 3946, 1969.
2. **Nunn, T. A.,** Miniature Absolute Pressure Transducer Assembly and Method, U. S. Patent 3,918,019, November 4, 1975.

# MINIATURE PRESSURE TRANSDUCER FOR BIOMEDICAL APPLICATIONS

**Jaroslav Hynecek**

Design development and pilot production of a miniature pressure transducer for biomedical application are described. The sensor is built on a single crystal silicon chip using standard methods of integrated circuit fabrication. The hermetically sealed vacuum chamber is used as a reference making an absolute pressure reading possible. Emphasis is given on testing procedures in order to fully characterize the fabricated device. A detailed description of the interface electronic circuitry for signal processing and temperature compensation is also included. Finally, the problem of packaging of the sensor for biomedical/body/environment is addressed; however, its complete solution is not as yet obtained.

## INTRODUCTION

A number of small pressure transducers intended for medical applications have recently been developed.[1–3] However, there are still several factors which have not been satisfactorily resolved and which are significantly restricting the use of these pressure transducers. The factors can be listed as follows: long-term base-line stability; temperature sensitivity; generally small pressure sensitivity which necessitates further amplification; hysteresis in both pressure and temperature cycle; poor resistance to the corrosive ambients such as body fluids; poor biocompatibility in general; and finally, excessively high cost.

In this report we present a design of a miniature pressure transducer which is aimed at minimizing some of the above-mentioned problems, namely hysteresis, long-term drift, biocompatibility, and optimizing the sensitivity. Unfortunately, not all of the problems have yet been resolved by this design; for example, the sensor still needs the temperature compensation. However we think that the data and results obtained during the development work are useful to other designers and manufacturers of the pressure transducers, and the sensor itself is useful to researchers in the biomedical field.

## DESCRIPTION OF THE SENSOR

**A. General** – In order to obtain low hysteresis of a pressure transducer, it is desirable to build it from only one material if possible. The silicon crystal is a natural choice for several reasons. The material is biocompatible.[4] Effect of the pressure or strain on the material electrical properties, piezoresistance in particular, is very well understood, and silicon technology is a highly developed art which allows one to incorporate a variety of different devices, sensing as well as amplifying or otherwise signal processing, right on the sensor itself. With this in mind, two basic mechanical structures were selected for construction of the sensor as it is pictured in Figures 1 and 2. The common part of both of these structures is a silicon chip of dimensions 50 × 150 mil and 5 mil thick, which has a preferentially etched cavity from the bottom of the chip in order to form a thin diaphragm on the top of the chip. The sensing elements are fabricated on the top and located in the center of the diaphragm. The shape and dimensions of the diaphragm are given in Figure 3. It is important to note that the mechanical properties of the silicon are anisotropic, as can be seen from the plot of the Poisson's ratio for the different crystallographic direction given in Figure 4.[5] The geometrical shape of the device therefore must be conformed to the crystal anisotropy in order to avoid unwanted stresses resulting from possible thermal expansion coefficient mismatch, if different materials are used for the diaphragm and package or supporting structure. A structure which is commonly used[6] and has this problem is a circular diaphragm mounted on the glass tube. The bending of the diaphragm in such a case cannot be simply described; anisotropy is usually not considered, and the reaction of the supporting structure is then unknown and undetermined. The rectangular shape of the diaphragm is therefore a natural choice which, in addition, can be easily fabricated by preferential chemical etching. The

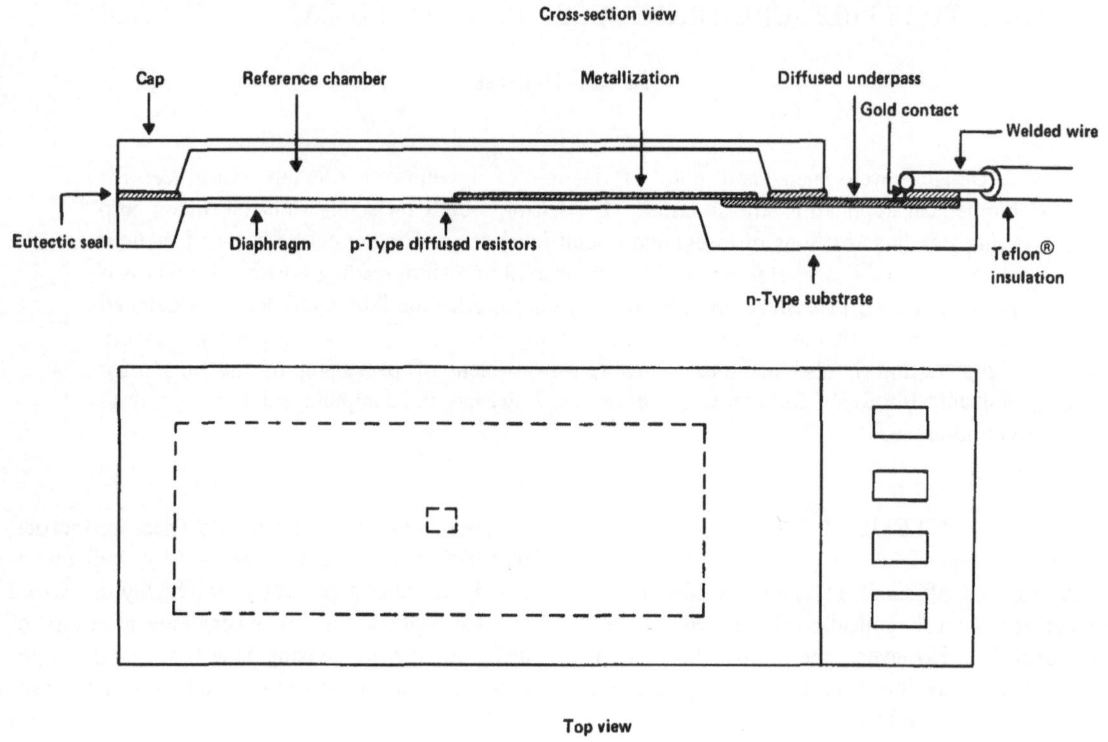


FIGURE 1. Pressure sensor structure 1.

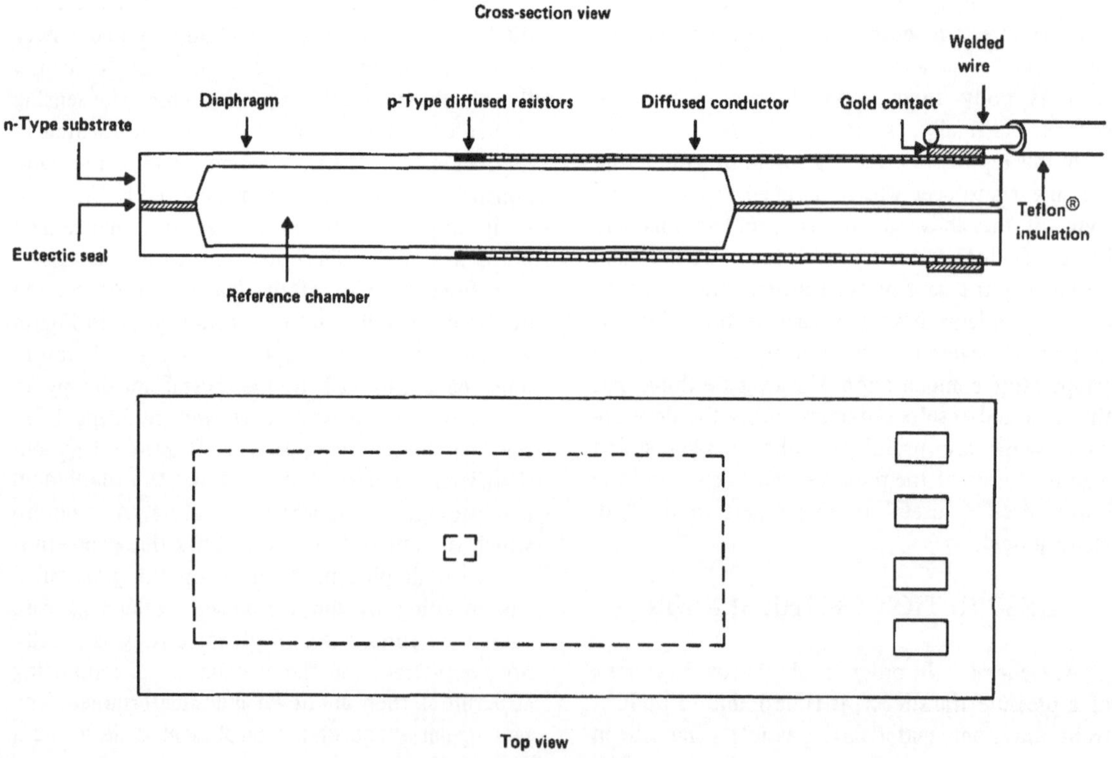


FIGURE 2. Pressure sensor structure 2.

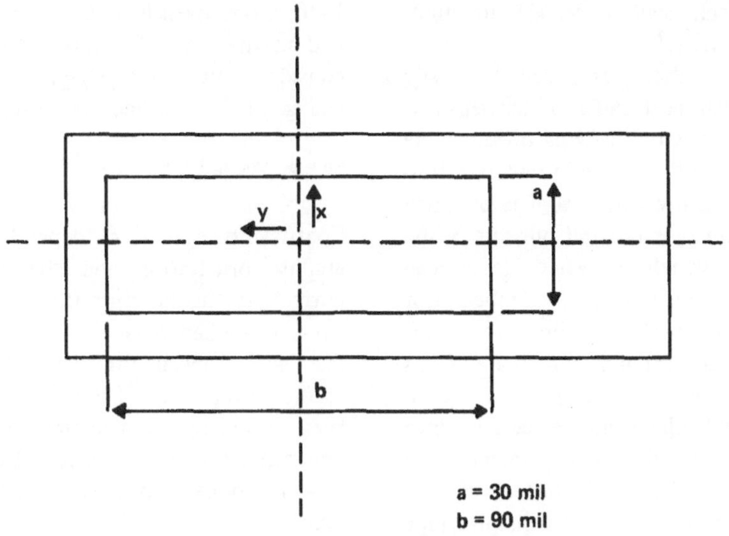


FIGURE 3. Dimensions of the diaphragm.

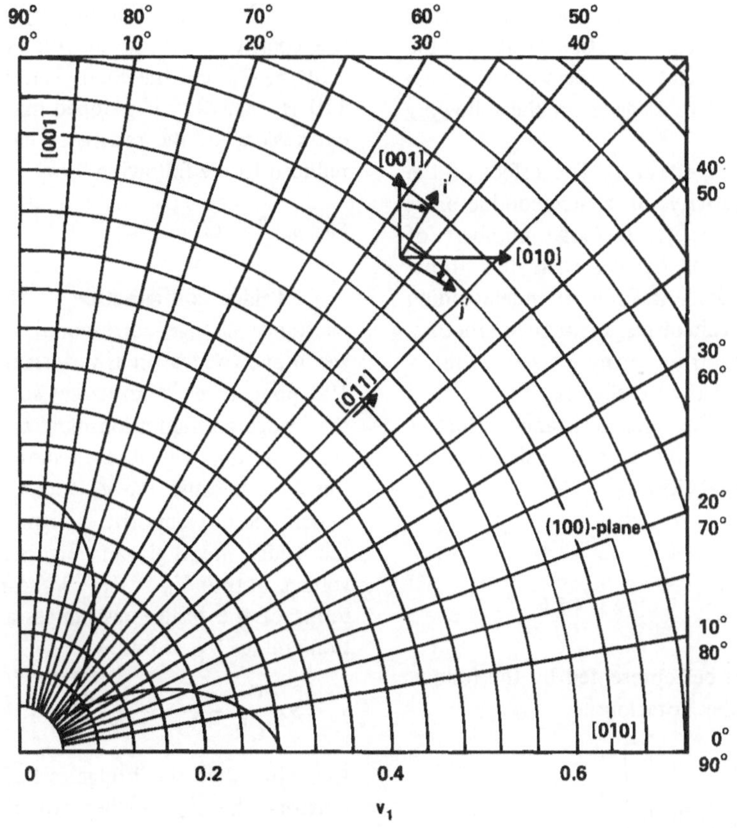


FIGURE 4. Poisson's ratio as a function of direction in the (100)-plane. (From Integrated Silicon Device Technology, Vol. V, AF 33(657)-10360, Research Triangle Institute, Durham, North Carolina, 1964, 11.)

etching rates are enhanced in certain directions and suppressed in others.[7]

The function of the sensor can be easily understood from Figures 1 and 2. If the sensor is placed in an ambient with changing pressure, the diaphragm deforms as a result of pressure differential load. The pressure differential is obtained between the ambient pressure and pressure in the reference chamber which is sealed. It is also possible to evacuate the reference chamber prior to sealing and obtain an absolute pressure sensor. Most of the sensors fabricated in our lab are of this type. The deformation of the diaphragm is then detected by a suitable electronic device built into it, which changes its electrical parameters in relation to the mechanical stresses in the diaphragm. In this design, a simple resistor bridge was fabricated on the diaphragm which produces output voltage proportional to the applied pressure.

In order to analyze the function of the sensor in more detail and to calculate the sensitivity, the mechanical as well as electrical aspect of the design must be discussed. This is given in the following sections.

**B. Piezoresistance effect** – The effect of the stress on the resistivity of a semiconductor is called piezoresistance effect. Large numbers of literature articles exist dealing with this topic, since it plays a fundamental role in understanding the conduction mechanisms of semiconductors. We list only a few from many important references; those which can be easily used for practical sensor design.[8–11] The piezoresistance effect is completely described by a piezoresistance coefficient tensor given below:

$$\rho_\alpha = \rho_0 \, [1 + \sum_{\beta=1}^{6} \pi_{\alpha\beta} \, \sigma_\beta] \qquad \alpha = 1,2,3$$
$$\rho_\alpha = \rho_0 \, [\sum_{\beta=1}^{6} \pi_{\alpha\beta} \, \sigma_\beta] \qquad \alpha = 4,5,6 \qquad (1)$$

The tensor $\pi_{\alpha\beta}$ can be represented by the following matrix in six-index notation:

$$\pi_{\alpha\beta} = \begin{bmatrix} \pi_{11} & \pi_{12} & \pi_{12} & 0 & 0 & 0 \\ \pi_{12} & \pi_{11} & \pi_{12} & 0 & 0 & 0 \\ \pi_{12} & \pi_{12} & \pi_{11} & 0 & 0 & 0 \\ 0 & 0 & 0 & \pi_{44} & 0 & 0 \\ 0 & 0 & 0 & 0 & \pi_{44} & 0 \\ 0 & 0 & 0 & 0 & 0 & \pi_{44} \end{bmatrix} \qquad (2)$$

Expressions given in Equation 1 can be simplified, and in the case of a thin diaphragm where we consider only lateral stresses, we can write the change of the resistance in the following form:

$$\delta\rho = \rho_0 \, (\pi_\parallel \, \sigma_\parallel + \tau \, \sigma_\perp) \qquad (3)$$

Coefficients $\pi_\parallel$ and $\pi_\perp$ depend on the crystallographic orientation and the direction of the current. In this notation the sign "∥" means that stress is applied parallel to the current flow, and the sign "⊥" means that the stress is perpendicular to the current flow. The values for $\pi_\parallel$ and $\pi_\perp$ have been evaluated for a number of crystallographic orientations as well as different doping types and concentrations, and they are given in Reference 10.

Since the chemical etching process was selected as a technique to fabricate the diaphragm, the choice of the crystal orientation is narrowed down to <100> silicon. For this crystallographic plane the coefficient $\pi_\parallel$ in the direction $1\bar{1}0$ has the value $\pi_{44}/2$ and the coefficient $\pi_\perp$ in the direction 110 is $-\pi_{44}/2$[10] if p-doped material is used. The expression for the resistivity change can then be reduced to the following form:

$$\delta\rho = \rho_0 \, \frac{\pi_{44}}{2} \, (\sigma_\parallel - \sigma_\perp) \qquad (4)$$

**C. Bridge configuration** – The amount of the resistance change corresponding to the pressures we intend to measure is very small, on the order of 100 ppm. In order to detect such a small change, the bridge configuration must be used. The bridge configuration will be even more advantageous in this case because the $\pi_\perp$ and $\pi_\parallel$ are the same in magnitude but opposite in signs. Considering now the bridge circuit given in Figure 5, the differential voltage output $V_0 = V_1 - V_2$ can be expressed in terms of the bridge resistances and bridge excitation voltage $V_B$ as follows:

$$V_0 = V_B \left[ \frac{R_3}{R_1 + R_3} - \frac{R_4}{R_2 + R_1} \right] \qquad (5)$$

Assuming that the bridge is built from identical resistors, the $V_0$ will be zero. For the change of the output voltage if we consider that $\delta R_3 = \delta R_2 = -\delta R_1 = -\delta R_4 = \delta R$, we can derive from Equation 5 the following relation:

$$\delta \, V_0 = V_B \frac{1}{R} \, \delta \, R \qquad (6)$$

where we assumed that $R_1 = R_2 = R_3 = R_4 = R$.

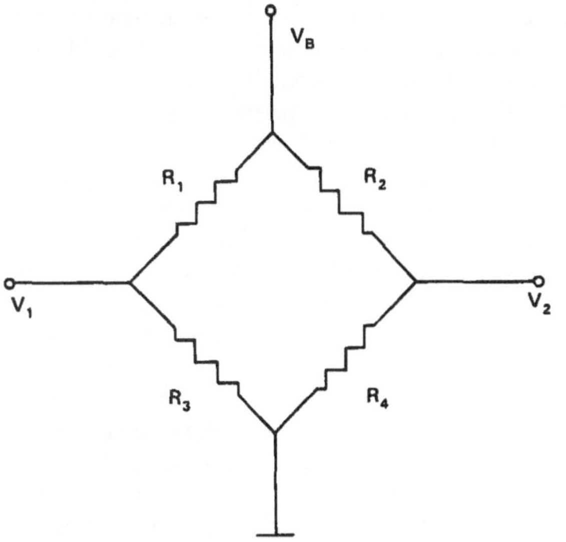


FIGURE 5. Bridge circuit.

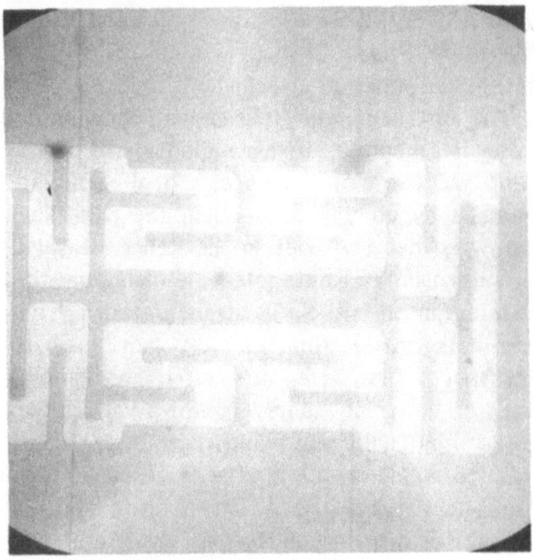

FIGURE 6. Photomicrograph of the pressure-sensing bridge diffused into the diaphragm.

Equation 6 will be used later in calculation of the sensitivity of the transducer.

**D. Mechanical considerations** – It is shown in Equation 4 that the change of the resistance is proportional to the difference in the longitudinal and transversal stress. It is desirable then to select such a mechanical structure that would enhance this difference. If, for example, a center of a circular diaphragm is used as a location for the piezoresistors, there would be no change in the resistance since $\delta_{\parallel} = \delta_R = \delta_{\perp} = \delta_T$. But for a center of a rectangular structure, Figure 3, it can be shown[13] that

$$\sigma_x = \frac{1}{4} P\left(\frac{a}{h}\right)^2 (1 - \nu), \quad \sigma_y = 0 \tag{7}$$

with the assumption that b > 3a. Several other different shapes for the diaphragm are currently under investigation, which can generate stresses of opposite signs and therefore increase the overall sensitivity of the transducer. Positioning the sensing elements at the edge of the diaphragm, where the stress is doubled due to the reaction of the support, was also considered, but it was found that due to the deformation of the support the stress was lower than expected and the sensor performance was furthermore very sensitive to the alignment of the chamber and etched cavity. The final design and location of the piezoresistors can be seen from the microphotograph in Figure 6.

**E. Pressure sensitivity** – The sensitivity of a transducer can be defined as a change of the output voltage with application of the pressure. Since the bridge configuration is used, the output voltage is also proportional to the bridge excitation voltage, and it is customary to normalize it to the bridge excitation voltage. The sensitivity then can be defined according to the following relation:

$$S = \frac{\delta V_o}{V_B} / \delta P \tag{8}$$

where $\delta P$ is change in the pressure load. By combining Equations 4, 6, 7, and 8 we obtain the following expression for the sensitivity:

$$S = \frac{1}{8} \pi_{44} \left(\frac{a}{h}\right)^2 (1 - \nu) \tag{9}$$

In deriving Equation 9 it was assumed that the stress is uniform in the area of the resistor location. This assumption requires that resistors are made small in dimensions with respect to the diaphragm width; otherwise, some cancellations of sensitivity will result. If the assumption is not satisfied, elaborate integration of the sensitivity over the length of the resistors has to be performed in order to obtain an exact formula. The expression in Equation 9, however, seems to be a satisfactory approximation for our case and most of the practical similar designs.

## FABRICATION PROCEDURE

The starting material used for fabrication is n-type <100> silicon with resistivity of 1 to 4

Ω·cm. The material was purchased from "Toyo"®[12] and is polished on both sides with the thickness of wafers approximately 5 mil.

The first step prior to standard circuit processing is generation of registration marks on the front and back of the wafer. These marks are obtained by etching windows in the oxide and regrowing the new oxide in the etched areas. All the subsequent processing steps are then aligned to these alignment marks. To assure a good registration of the marks from the back of the wafer to the front of the wafer, the wafers are photoresisted on both sides and exposed in a jig with two masks, front and back, which have been aligned prior to wafer insertion. The subsequent processing steps have two goals. The first is to fabricate a low resistivity region serving as underpassing leads to bring the bridge signal out from the reference chamber through the vacuum seal (Figure 1) and also provide good bonding pads for future wire attachment. The second is to fabricate the piezoresistor bridge in the center of the diaphragm. The standard processing sequence is used for fabrication, resulting in resistors with values from 4 to 5 kΩ and surface concentration of $C_s = 10^{18} cm^{-3}$. Choice of this surface concentration is a compromise between temperature drift and lower sensor pressure sensitivity.[10] There is a small difference in design between the sensor in Figure 1 and the sensor in Figure 2. The first type has diffused underpasses only under the vacuum seal and bonding pads, while metallization fills in the rest of the connections. The second type of the transducer (Figure 2) has diffused leads all the way from the resistor bridge to the bonding pads. Elimination of the metallization sections over the diaphragm was aimed to reduce hysteresis resulting from a plastic deformation of the metal. After the diffusion processing steps are completed, the metallization is performed by vacuum deposition and a chrome-gold combination is used. The metals are evaporated from the tungsten boats, and wafers are heated to 250°C. Both sides of the wafers are metallized. The front is metallized over the passivation oxide, whereas the oxide on the back is stripped off and metal is directly in contact with silicon. After metallization, the wafers are gold-plated in order to provide enough metal thickness for alloy seal of the reference chamber and for the wire attachment. It was found experimentally that thickness over 2 μm is usually sufficient to obtain a good vacuum-tight seal. The next operation which follows is a chemical preferential etch of the silicon.[7] The thickness of the diaphragm fabricated in this way is usually between 0.5 to 1.0 mil, and it is possible to maintain a 0.25-mil tolerance. The etching is performed in a heated flask equipped with a reflux condenser in order to maintain a constant concentration of the solution during the etching. The last operation of the pressure-sensor chip manufacturing is the vacuum sealing of the reference chamber. An eutectic gold-tin alloy preform which is used in seal is manufactured by sandblasting through a stainless steel mask. This procedure was selected because it was difficult to etch the gold-tin alloy directly. The actual seal is done in a vacuum chamber where assembled chip with preform and either reference chamber cup or bottom chip (Figure 2) are heated through the temperature cycle. A proper temperature cycle was found experimentally in order to obtain a good seal with minimum stress which could have undesirable effects on the hysteresis of the sensor. This step completes manufacturing of the pressure-sensing chip, which is then temporarily mounted in a flat-pack for testing purposes before it goes into the final packaging step. The temperature cycle data for the design given in Figures 1 and 2 are given in Figures 7 and 8 for comparison.

## TESTING AND PACKAGING

The testing procedure includes both pressure and temperature cycling. The transducer is first pressure-cycled several times to measure its sensitivity and also any pressure hysteresis. Next the transducer is temperature-cycled to determine temperature coefficient and hysteresis. Since the output signal from the transducer is small, on the order of 10 μV/mmHg, it was necessary to build low-drift amplifiers (Figure 9) to perform the desired measurements. The amplifiers include an offset adjustment to compensate variations between sensors. The temperature data are obtained from a thermocouple using an ice-bath reference junction. The pressure transducer and thermocouple signals are both amplified and used to drive an x-y plotter and automatic temperature-cycling apparatus.

The automatic temperature unit (Figure 10) controls the flow of hot and cold water to the test chamber. The chamber itself is a small aluminum vessel, with air circulation fan, which is mounted on a thermoelectric cooler.

During the cooling cycle, the thermoelectric

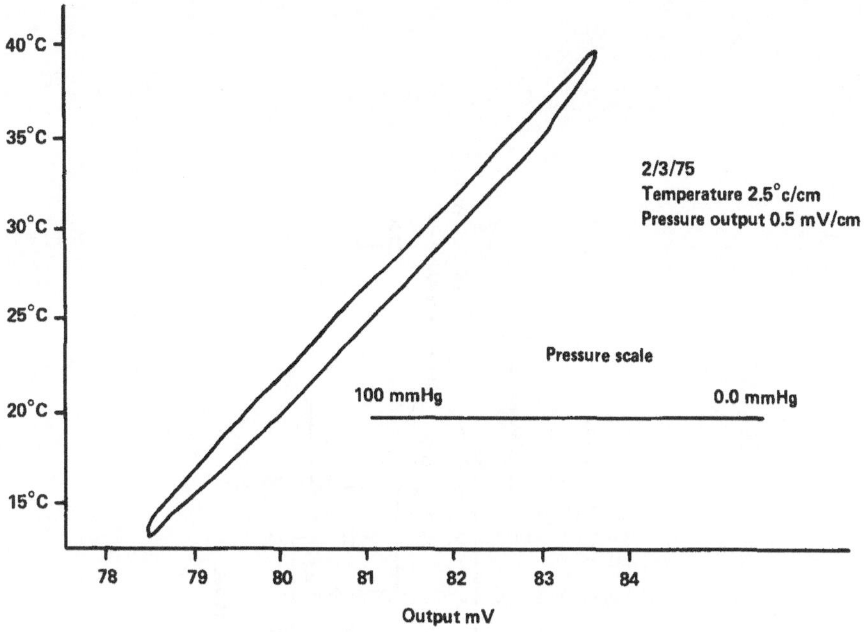


FIGURE 7. Performance of sensor 3.

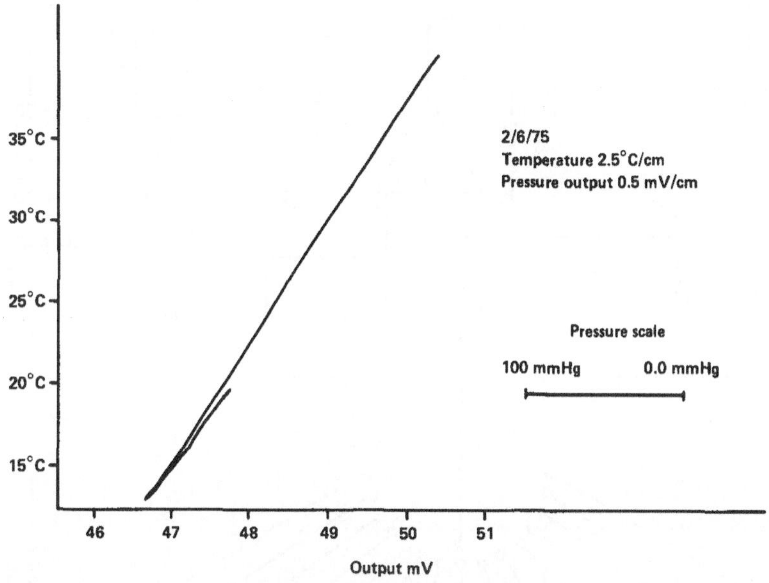


FIGURE 8. Performance of sensor 10.

cooler is turned on and cold water is passed over the junction as a heat sink. During the heating cycle, the cooler is switched off and the control unit switches the water flow to a heated supply to warm the junction. The rate of cycling is determined by the water flow rate and the thermocooler current, while the cycle limits are part of the automatic control.

Testing is done on the transducer prior to packaging in order to eliminate low performance units and also to evaluate the packaging methods by giving comparative reference data.

A typical temperature cycle curve is given in Figure 8. At this point the transducer is mounted in a flat-pack and can be used as an absolute pressure-sensing device in gaseous ambients or in

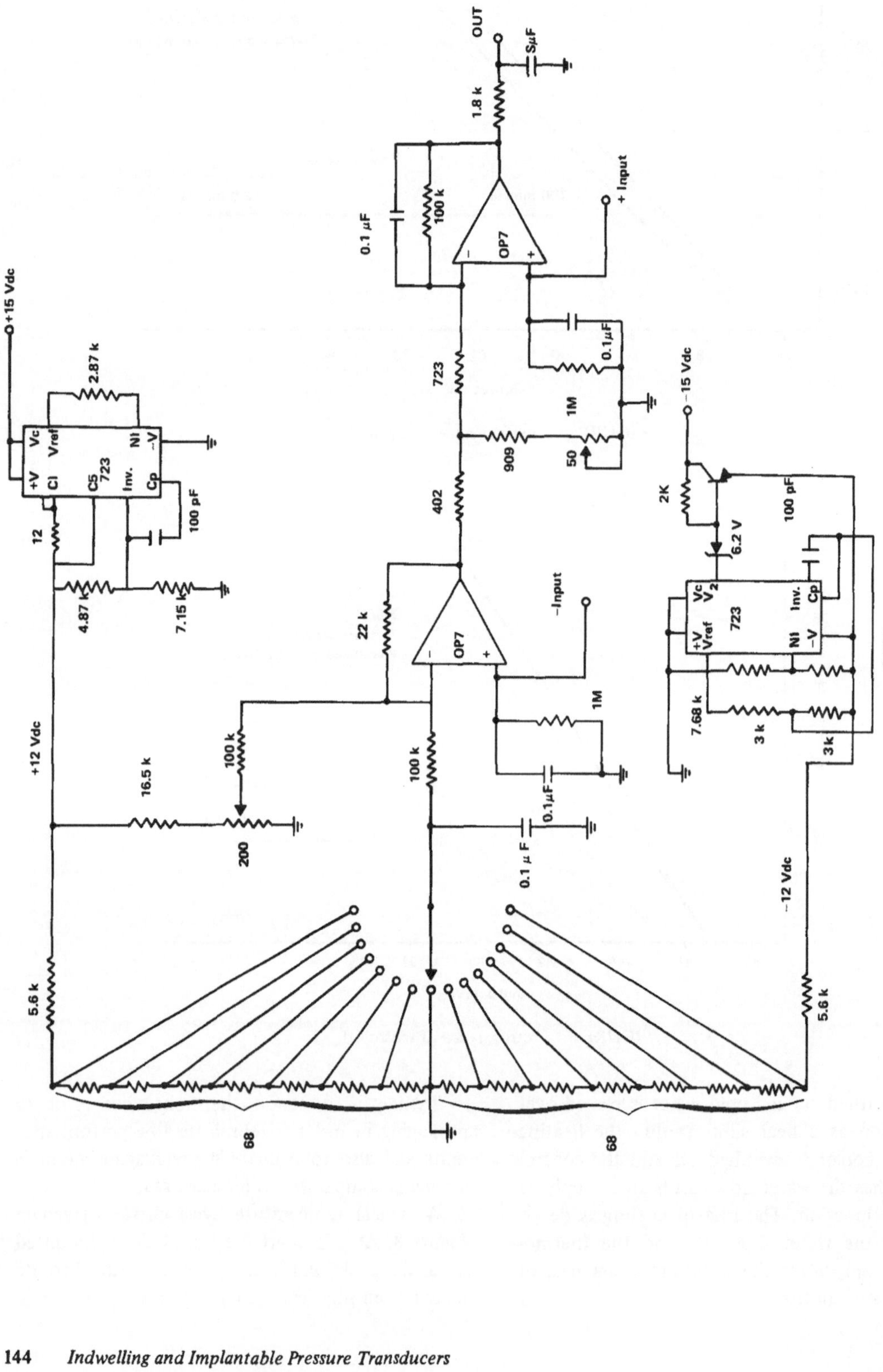


FIGURE 9. Low-drift amplifier circuit diagram.

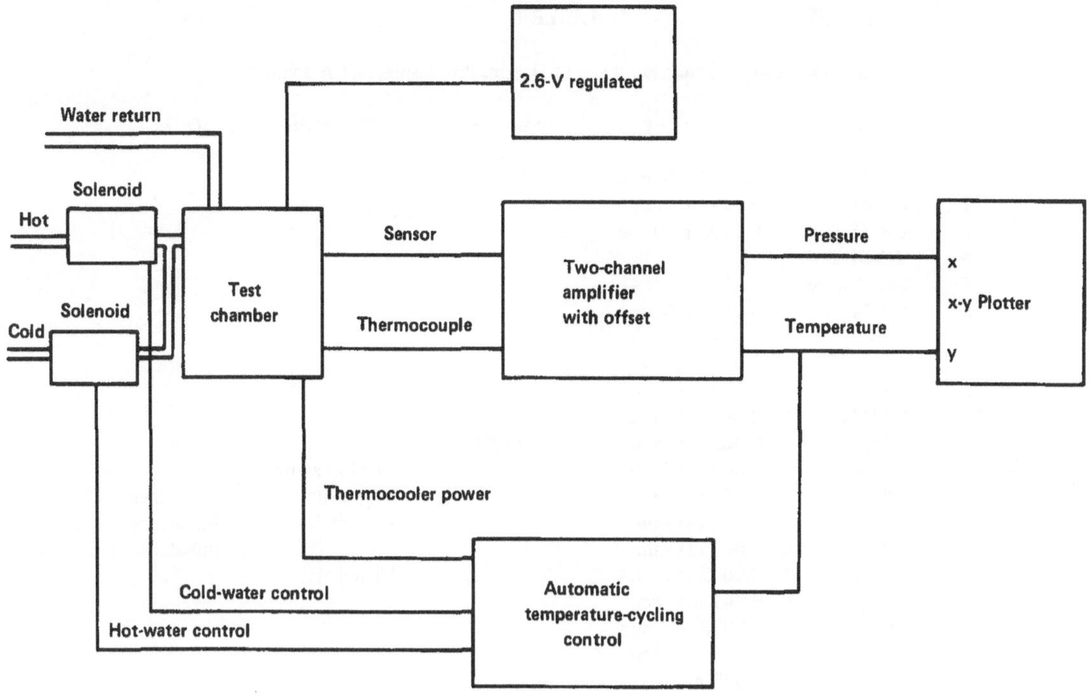


FIGURE 10. Temperature testing setup block diagram.

nonconductive liquid ambient. With the addition of the electronic compensation amplifiers developed for the transducer, stability can be maintained to within 1 mmHg between room temperature and 40°C.

At this time almost 200 transducers have been tested. Based on the data obtained from early transducers, several evolutionary improvements in the design have been developed. Simultaneously, work on electronic amplification and compensation, as well as packaging, has moved through several phases. The tables and graphs that follow show data on transducers with and without packaging and/or compensation.

An initial group of 35 transducers was constructed as shown in Table 1. This group was used to study the feasibility of the design and the problems that processing and assembly would present. These problems included imbalance in the bridge, drift of the operating point, and leakage in the seal of the transducer. Of these transducers, six were unbalanced enough to be judged useless.

The average sensitivity of the transducers in this group was 0.028 mV/mm Hg, with variations from 0.0038 mV/mmHg to 0.112 mV/mmHg. In this first group, the problems of temperature drift and hysteresis became apparent. The amount of data taken on variables was small in this group; however, temperature coefficients were on the order of -6 mm/°C, and hysteresis was on the order of 10 to 20 mm at the center of the temperature cycle used (25°C). This group of sensors was sealed with epoxy which was replaced by an eutectic solder seal in the later sensors to reduce temperature effects in order to provide a more durable seal.

In order to eutectically seal the transducer diaphragm and cap, it is necessary to heat these pieces, with the gold-tin solder preform between them, to 280°C. At this temperature it was felt that in an oxygen atmosphere the metallization and solder would be attacked, therefore two sensors were experimentally sealed eutectically in an inert argon atmosphere. The data on these sensors are shown in Table 2.

The initial group of transducers was sealed in air at atmospheric pressure. Consideration of this factor suggested that the air trapped inside the reference chamber expanded with temperature and increased the temperature drift of the devices. It was decided to seal the sensors in a vacuum, which also eliminated any oxidation problems. A group of transducers, designated VS, was assembled using the new diaphragms designed for eutectic sealing. The data for this group are shown in Table 3. In

TABLE 1

**Sensors Used to Study Problems of Design, Processing, and Assembly**

| No. | Date | Sensitivity | Temp coeff | Hysteresis | Offset |
|---|---|---|---|---|---|
| 1 | 1/8/74 | 0.0114 mV/mm | | | |
| 2 | 1/22/74 | 0.024 mV/mm | | | |
| 3 | 1/8/74 | 0.0195 mV/mm | | | |
| 4 | Unbalanced | | | | |
| 5 | Unbalanced | | | | |
| 6 | 1/8/74 | 0.0139 mV/mm | | | |
| 7 | 1/10/74 | 0.011 mV/mm | | | |
| 8 | 1/11/74 | 0.031 mV/mm | | | |
| 9 | 1/14/74 | 0.033 mV/mm | | | |
| 10 | 1/14/74 | 0.047 mV/mm | | | |
| 11 | 3/15/74 | 0.027 mV/mm | −11 mm/°C | | |
| 12 | 3/7/74 | 0.002 mV/mm | | Bad imbalance | |
| 13 | 3/8/74 | 0.04 mV/mm | | High drift | Imbalance |
| 14 | 3/8/74 | 0.009 mV/mm | | High drift | Imbalance |
| 15 | 3/8/74 | 0.011 mV/mm | | | Imbalance |
| 16 | 3/11/74 | 0.0042 mV/mm | | High drift | |
| 17 | 3/11/74 | 0.041 mV/mm | | | |
| 18 | 3/12/74 | 0.0038 mV/mm | | | |
| 19 | 3/12/74 | 0.0085 mV/mm | | | |
| 20 | 4/3/74 | 0.028 mV/mm | | | |
| 21 | 4/3/74 | 0.0135 mV/mm | | | |
| 22 | | | | Leaks | |
| 23 | 4/3/74 | 0.028 mV/mm | −5.3 mm/°C | 83 mm | |
| 24 | | | | Leaks | |
| 25 | 4/19/74 | 0.0094 mV/mm | −8 mm/°C | 3.5 mm | |
| 26 | 4/18/74 | 0.02 mV/mm | −4 mm/°C | 15 mm | |
| 27 | 4/19/74 | 0.014 mV/mm | | | |
| 28 | 4/19/74 | 0.021 mV/mm | | | |
| 29 | 4/18/74 | 0.112 mV/mm | −2.85 mm/°C | 9 mm | |
| 30 | 4/2/74 | 0.053 mV/mm | | Leaks | |
| 31 | 4/2/74 | 0.026 mV/mm | | Leaks | |
| 32 | 4/11/74 | 0.062 mV/mm | −53.5 mm/°C | 48 mm | |
| 33 | 4/5/74 | 0.1 mV/mm | | Leaks | |
| 34 | 4/8/74 | 0.02 mV/mm | −7.5 mm/°C | 50 mm | |
| 35 | 4/8/74 | 0.015 mV/mm | −1.65 mm/°C | 20 mm | |

TABLE 2

**Argon-filled Sensors Designated AR**

| No. | Date | Sensitivity | Temp coeff | Hysteresis |
|---|---|---|---|---|
| AR-1 | 10/30/74 | 0.027 mV/mm | −11.2 mm/°C | 11 mm |
| AR-2 | 10/31/74 | 0.019 mV/mm | −7.2 mm/°C | 23 mm |

this group the average sensitivity was 0.042 mV/mmHg, the temperature coefficient of drift was −6.37 mmHg/°C, and hysteresis was 5.52 mmHg. The average imbalance offset for this group was 73.5 mV.

At this time a small number of diaphragms (Table 4) were tested to determine if an assembled transducer had the same temperature coefficient as a diaphragm. The diaphragms had a temperature coefficient of 0.038 mV/°C, while the assembled transducers had an equivalent coefficient of 0.27 mV/°C. It was therefore concluded that the strain induced in a finished sensor creates a significant increase in the temperature coefficient. Since this temperature coefficient cannot be removed by preprocessing techniques, some form of compensation is necessary for any transducer.

Also at this time a group of transducers

TABLE 3

**Eutectic-sealed Sensors Designated VS**

| No. | Date | Sensitivity | Temp coeff | Hysteresis Top | Bottom | Spread | Offset |
|---|---|---|---|---|---|---|---|
| VS-1 | 1/15/75 | 0.037 mV/mm | -6.5 mm/°C | 19° | 9° | 8.8 mm | 63.9 mV |
| 2 | 1/16/75 | 0.048 mV/mm | -5.3 mm/°C | 1° | 1° | 0.53 mm | 85 mV Drift |
| 3 | Leaks | | | | | | |
| 4 | 1/16/75 | 0.074 mV/mm | -2.7 mm/°C | 3° | 2° | 1.37 mm | 94.9 mV |
| | 1/23/75 | | -2.5 mm/°C | 1° | 1° | 0.34 mm | Low-temp cycle |
| 5 | | | | | | | Long-term |
| 6 | 1/8/75 | 0.073 mV/mm | -4.2 mm/°C | 2° | 2° | 1.4 mm | 68.5 mV |
| 7 | | | | | | | Broken |
| 8 | | | | | | | Packaging |
| 9 | 1/7/75 | 0.018 mV/mm | -11.6 mm/°C | 20° | 20° | 1.4 mm | 68.5 mV |
| 10 | 1/17/75 | 0.017 mV/mm | -8.7 mm/°C | 2° | 3° | 7.3 mm | 57.8 mV |
| 11 | | | | | | | Leaks |
| 12 | 1/20/75 | 0.061 mV/mm | -4.8 mm/°C | 5° | 1° | 0.4 mm | 89.5 mV |
| 13 | | | | | | | Leaks |
| 14 | 1/8/75 | 0.024 mV/mm | -8.2 mm/°C | 10° | 22° | 28.4 mm | |
| | 1/23/75 | | -9.3 mm/°C | 7° | 7° | 7.9 mm | Low-temp cycle |
| 15 | | | | | | | Leaks |
| 16 | 1/20/75 | 0.04 mV/mm | -4.45 mm/°C | 7° | 5° | 2.4 mm | 59.1 mV |
| 17 | 1/6/75 | 0.02 mV/mm | -4.3 mm/°C | 9° | 9° | 10 mm | |
| 18 | 2/6/75 | 0.030 mV/mm | -10.1 mm/°C | 2° | 2° | 2.7 mm | 52.8 mV |
| 19 | 2/6/75 | 0.065 mV/mm | -5.6 mm/°C | 11° | 11° | 1.5 mm | 79.6 mV |
| Average value: | | 0.042 mV/mm | -6.37 mm/°C | | | 5.52 mm | 73.5 mV |

TABLE 4

**Diaphragm Tests**

| No. | Date | Temp coeff | Offset |
|---|---|---|---|
| 1 | 11/25/74 | 0.066 mV/°C | |
| 2 | 11/22/74 | 0.025 mV/°C | |
| 3 | 11/21/74 | 0.020 mV/°C | 32.2 mV |
| 4 | 11/20/74 | 0.024 mV/°C | 35.0 mV |
| 5 | 12/1/74 | 0.021 mV/°C | 51.3 mV |
| 6 | 12/3/74 | 0.070 mV/°C | 57.8 mV |
| Average value: | | 0.038 mV/°C | |

designated BB was assembled using two diaphragms back to back instead of a diaphragm and a cap (Table 5). It was hoped that this would equalize the stress in the transducer and remove the temperature drift. The average temperature coefficient for this group was 1.8 mmHg/°C. This design lowered the temperature coefficient of the transducers as hoped; however it left the metallization on the sensors exposed, which increased the packaging problem.

From the data taken on temperature cycling, it appeared that the temperature hysteresis spread was dependent on the range of the temperature cycle. Several transducers were cycled over a lower temperature range (10°C) to test this. Over this small temperature range the average hysteresis spread was increased by 81%. This suggests that the sensor would give extremely good results in an implantation application environment where temperature variation would be limited.

It was decided at this point to test the diaphragms of each transducer prior to assembling the complete devices. Forty-eight diaphragms were tested in this way and 40 sensors assembled and designated CNC (Table 6). This group of transducers had an average sensitivity of 0.0566 mV/mmHg, a temperature coefficient of -4.66 mmHg/°C, and hysteresis spread of 5.9 mmHg. Offsets for this group were typically 50 mV. For this group it was again noted that the temperature coefficient of the assembled sensor was 0.27 mV/°C or almost ten times that of the diaphragm. Also apparent from this data was the fact that in most cases the diaphragm itself had little or no hysteresis compared to the assembled transducer, and that the hysteresis is therefore introduced in the assembly process.

TABLE 5

**Epoxy-sealed Back-to-back Sensors Designated BB**

| No. | Date | Sensitivity | Temp coeff | Hysteresis Top | Hysteresis Bottom | Hysteresis Spread | Offset |
|---|---|---|---|---|---|---|---|
| BB-1 | Leaks | | | | | | |
| 2 | 1/14/75 | 0.06 mV/mm | −1.94 mm/°C | 3° | 5° | 3.4 mm | 37.8 mV |
| 3 | 1/9/75 | 0.047 mV/mm | −1.59 mm/°C | 23° | 11° | 7.0 mm | |
| 4 | 1/9/75 | 0.072 mV/mm | −2.4 mm/°C | 70° | 23° | 7.9 mm | 40.9 mV |
| | 1/23/75 | | −1.6 mm/°C | 2° | 2° | 0.63 mm | Low-temp cycle |
| 5 | 1/10/75 | 0.126 mV/mm | −2.3 mm/°C | 94° | 50° | 6.9 mm | 57.7 mV |
| | 1/24/75 | | −2.0 mm/°C | 2° | 2° | 0.8 mm | Low-temp cycle |
| 6 | 1/13/75 | 0.052 mV/mm | −1.83 mm/°C | 33° | 16° | 5.7 mm | 44 mV |
| 7 | 1/13/75 | 0.039 mV/mm | −0.82 mm/°C | 20° | 5° | 4.5 mm | 37.9 mV |
| 8 | 1/14/75 | 0.069 mV/mm | −2.4 mm/°C | 8° | 7° | 3.3 mm | 56.1 mV |
| Average value: | | 0.066 mV/mm | 1.88 mm/°C | | | | |

TABLE 6

**Sensor Diaphragms Designated NC and Completed Sensors Designated CNC**

| No. | Date | Sensitivity | Temp coeff | Hysteresis Top | Hysteresis Bottom | Hysteresis Spread | Offset |
|---|---|---|---|---|---|---|---|
| NC-1 | 1/21/75 | | −0.023 mV/°C | | 0 | | |
| NC-2 | 1/21/75 | | −0.023 mV/°C | | 0 | | 43.6 mV |
| NC-3 | 1/21/75 | | −0.018 mV/°C | | 0 | | |
| CNC-3 | 2/3/75 | 0.04 mV/mm | −0.84 mm/°C | 34° | 3° | 10 mm | 80.4 mV |
| NC-4 | 1/10/75 | Leaks | −0.025 mV/°C | 1° | 1° | 0.025 mV | 39.4 mV |
| CNC-4 | 2/4/75 | | −0.1 mV/°C | 2° | 2° | 0.05 mV | 56.4 mV |
| NC-5 | 1/21/75 | | −0.019 mV/°C | | 0 | | 50.6 mV |
| CNC-5 | 2/4/75 | 0.05 mV/mm | −4 mm/°C | 16° | 18° | 5 mm | 93.6 mV |
| NC-6 | 1/22/75 | | −0.068 mV/°C | | 0 | | 56.7 mV |
| CNC-6 | 2/4/75 | Leaks | −0.148 mV/°C | 5° | 5° | 0.125 mV | 73.6 mV |
| NC-7 | 1/27/75 | | −0.05 mV/°C | | 0 | | 13.2 mV |
| CNC-7 | 2/5/75 | 0.073 mV/mm | −3.1 mm/°C | 12° | 12° | 2 mm | 71.8 mV |
| NC-8 | | | | | | | Unbalanced |
| NC-9 | 1/24/75 | | −0.063 mV/°C | 1° | 1° | 0.025 mV | 30.9 mV |
| CNC-9 | 2/5/75 | 0.022 mV/mm | −8.6 mm/°C | 2° | 2° | 6.9 mm | 45.9 mV |
| NC-10 | 1/28/75 | | −0.063 mV/°C | | 0 | | 25.8 mV |
| CNC-10 | 2/6/75 | 0.022 mV/mm | −6.2 mm/°C | 1° | 1° | 1.6 mm | 48.5 mV |
| NC-11 | 1/27/75 | | −0.10 mV/°C | 2° | 2° | 0.075 mV | −1.6 mV |
| CNC-11 | 2/7/75 | Leaks | −0.128 mV/°C | 2° | 2° | 0.045 mV | 4.6 mV |
| NC-12 | 1/28/75 | | 0.005 mV/°C | | 0 | | −2.8 mV |
| NC-13 | 1/29/75 | | −0.036 mV/°C | 1° | 1° | 0.025 mV | 53.4 mV |
| NC-14 | 1/29/75 | | −0.07 mV/°C | 1° | 1° | 0.25 mV | 24.2 mV |
| CNC-14 | 2/7/75 | Leaks | −0.12 mV/° | 4° | 4° | 0.045 mV | 33.2 mV |
| NC-15 | | | | | | | Noise |
| NC-16 | 1/29/75 | | −0.06 mV/mm | 1° | 1° | 0.025 mV | 26.2 mV |
| CNC-16 | 3/20/75 | 0.023 mV/mm | −7.1 mm/°C | 10° | 10° | 17.4 mm | 58.1 mV |
| NC-17 | 2/12/75 | | −0.043 mV/°C | | | 0 | 46.2 mV |
| CNC-17 | 3/17/75 | Leaks | −0.066 mV/°C | 1° | 1° | 0.025 mV | 53.6 mV |
| NC-18 | 2/12/75 | | −0.04 mV/°C | | 0 | | 37.0 mV |
| CNC-18 | 3/10/75 | 0.038 mV/mm | −3.57 mm/°C | 18° | 28° | 29 mm | 55.2 mV |
| NC-19 | 2/13/75 | | −0.09 mV/°C | 2° | 2° | 0.05 mV | 54.7 mV |
| CNC-19 | 3/10/75 | 0.055 mV/mm | −6.3 mm/°C | 2° | 2° | 1.4 mm | 154.0 mV |

TABLE 6 (continued)

**Sensor Diaphragms Designated NC and Completed Sensors Designated CNC**

| No. | Date | Sensitivity | Temp coeff | Hysteresis Top | Hysteresis Bottom | Hysteresis Spread | Offset |
|---|---|---|---|---|---|---|---|
| NC-20 | 2/13/75 | | -0.045 mV/°C | | 0 | | 46.4 mV |
| CNC-20 | 3/7/75 | 0.048 mV/mm | -4.16 mm/°C | 3° | 2° | 1.6 mm | 50.4 mV |
| NC-21 | 2/14/75 | | -0.057 mV/°C | | 0 | | 42.9 mV |
| NC-22 | 2/14/75 | | -0.053 mV/°C | 1° | 1° | 0.025 mV | 26.6 mV |
| NC-23 | 2/14/75 | | -0.067 mV/°C | 1° | 1° | 0.025 mV | 27.6 mV |
| NC-24 | | | | | | | Noise |
| NC-25 | 2/17/75 | | -0.145 mV/°C | 1° | 2° | 1.8 mV | 27.0 mV |
| CNC-25 | 3/14/75 | 0.041 mV/mm | -6.8 mm/°C | 2° | 2° | 1.8 mm | 27.0 mV |
| NC-26 | 2/17/75 | | -0.095 mV/°C | 1° | 1° | 0.025 mV | 11.0 mV |
| NC-27 | 2/17/75 | | -0.063 mV/°C | | 0 | | 35.1 mV |
| CNC-27 | 3/18/75 | 0.033 mV/mm | -5.2 mm/°C | 2° | 2° | 3 mm | 56.2 mV |
| NC-28 | 2/19/75 | | -0.05 mV/°C | | 0 | | 43.8 mV |
| CNC-28 | 3/31/75 | 0.032 mV/mm | -2.38 mm/°C | | 0 | | 49.6 mV |
| NC-29 | 3/19/75 | | -0.078 mV/°C | 2° | 2° | 0.05 mV | 19.2 mV |
| CNC-29 | 3/31/75 | 0.039 mV/mm | -4.6 mm/°C | 49° | 50° | 7.7 mm | 41.0 mV |
| NC-30 | 3/19/75 | | -0.046 mV/°C | | 0 | | 45.9 mV |
| CNC-30 | 3/31/75 | 0.058 mV/mm | -4.3 mm/°C | | 0 | | 95.9 mV |
| NC-31 | 3/20/75 | | -0.077 mV/°C | 1° | 1° | 0.025 mV | 105 mV |
| CNC-31 | 4/2/75 | 0.023 mV/mm | -7.5 mm/°C | 1° | 1° | 1.8 mm | 109 mV |
| NC-32 | 3/25/75 | | -0.056 mV/°C | | 0 | | 39.6 mV |
| CNC-32 | 4/11/75 | 0.056 mV/mm | -2.94 mm/°C | 1° | 1° | 0.44 mm | 60.9 mV |
| NC-33 | 3/25/75 | | -0.08 mV/°C | | 0 | | 5.1 mV |
| NC-34 | 3/28/75 | | -0.05 mV/°C | 1° | 1° | 0.025 mV | 45.9 mV |
| NC-35 | 3/27/75 | | -0.076 mV/°C | | 0 | | 15.2 mV |
| NC-36 | 3/27/75 | | -0.0025 mV/°C | | 0 | | 35.9 mV |
| CNC-36 | 5/9/75 | 0.017 mV/mm | -5.6 mm/°C | 2° | 3° | 3 mm | 49.9 mV |
| NC-37 | 3/28/75 | | -0.08 mV/°C | 5° | 5° | 0.05 mV | 51.9 mV |
| NC-38 | 4/3/75 | | 0.188 mV/°C | 0 | 0 | 0.025 mV | 60.4 mV |
| NC-39 | 4/3/75 | | -0.05 mV/°C | 6° | 5° | 0.15 mV | 75.3 mV |
| CNC-39 | 4/11/75 | 0.019 mV/mm | -1.75 mm/°C | | 0 | | 50.0 mV |
| NC-40 | 4/13/75 | | -0.01 mV/°C | | 0 | | 30.6 mV |
| CNC-40 | 4/21/75 | 0.026 mV/mm | -3.65 mm/°C | 2° | 2° | 2 mm | 50.2 mV |
| NC-41 | | | | | | | Unbalanced |
| NC-42 | 4/21/75 | | 0 | | | | 38.5 mV |
| NC-43 | 4/21/75 | | -0.035 mV/°C | 0 | 0 | 0.025 mV | 76.2 mV |
| NC-44 | 4/22/75 | | -0.0025 mV/°C | | 0 | | 53.8 mV |
| NC-45 | 4/22/75 | | -0.045 mV/°C | 2° | 2° | 0.05 mV | 44.7 mV |
| NC-46 | 4/22/75 | | 0.031 mV/°C | 1° | 1° | 0.025 mV | 69.4 mV |
| NC-47 | 4/22/75 | | -0.0025 mV/°C | | 0 | | 37.9 mV |
| NC-48 | 4/23/75 | | -0.134 mV/°C | 3° | 3° | 0.1 mV | 82.4 mV |
| Average value: | | 0.0566 mV/mm | -4.66 mm/°C | | | 5.9 mm | |

A second set of diaphragms, designated BDC, were processed using the experience of the last group to improve the tolerances of the devices (Table 7). This group of devices gave much less variation in all parameters. The average sensitivity of this group was 0.0223 mV/mmHg, the temperature coefficient was -6.5 mmHg/°C, and the hysteresis spread was 5.87 mmHg. The offset of this group was again typically 50 mV.

Based on the results of all the transducers tested up to this point, a new transducer was designed, designated CD (Table 8). In this device the bridge is placed on the outside of the sensor and protected by a gold metallization. This gold metallization completely covered the transducer except for 40 mil at the bonding pad end of the diaphragm. The sensor is assembled by bonding two diaphragms together back to back, as in the

TABLE 7

**Pressure Diaphragms Designated BD and Pressure Sensors Designated BDC**

| No. | Date | Sensitivity | Temp coef | Hysteresis Top | Hysteresis Bottom | Hysteresis Spread | Offset |
|---|---|---|---|---|---|---|---|
| BD-1 | 4/23/75 | | -0.02 mV/°C | 1° | 1° | 0.025 mV | 58.8 mV |
| BDC-1 | 4/29/75 | 0.02 mV/mm | -7.5 mm/°C | 3° | 2° | 3.75 mm | 69.3 mV |
| BD-2 | 4/23/75 | | -0.091 mV/mm | 0 | 2° | 0.05 mV | 50.8 mV |
| BDC-2 | 5/6/75 | 0.023 mV/mm | -4.84 mm/°C | 2° | 1° | 1 mm | 54.9 mV |
| BD-3 | 4/23/75 | | -0.025 mV/°C | 0 | 1° | 0.025 mV | 52.4 mV |
| BDC-3 | 6/25/75 | 0.019 mV/mm | -5.9 mm/°C | 2° | 2° | 2.6 mm | 73.7 mV |
| BD-4 | 4/24/75 | | -0.025 mV/°C | 0 | 2° | 0.05 mV | 45.0 mV |
| BDC-4 | 5/6/75 | 0.019 mV/mm | -9.3 mm/°C | 2° | 2° | 4 mm | 81.8 mV |
| BD-5 | 4/24/75 | | -0.021 mV/°C | 0 | 2° | 0.025 mV | 45.3 mV |
| BDC-5 | 6/25/75 | 0.02 mV/mm | -4 mm/°C | 5° | 8° | 5.2 mm | 40.6 mV |
| BDC-6 | 5/7/75 | 0.019 mV/mm | -5.2 mm/°C | 9° | 12° | 19.6 mm | 56.3 mV |
| BDC-7 | 5/7/75 | 0.02 mV/mm | -5.9 mm/°C | 3° | 3° | 9.7 mm | 53.1 mV |
| BDC-8 | 5/8/75 | 0.033 mV/mm | -3.4 mm/°C | 3° | 3° | 1.5 mm | 32.2 mV |
| BDC-9 | 5/8/75 | 0.026 mV/mm | -6.2 mm/°C | 3° | 2° | 2.0 mm | 64.7 mV |
| BDC-10 | 5/14/75 | 0.02 mV/mm | -2.7 mm/°C | 26° | 28° | 25.6 mm | 48.6 mV |
| BDC-11 | 5/14/75 | 0.02 mV/mm | -4.2 mm/°C | | 0 | | 51.9 mV |
| BDC-12 | 5/15/75 | 0.02 mV/mm | -6.4 mm/°C | 1° | 0 | 1.3 mm | 53.9 mV |
| BDC-13 | 5/15/75 | 0.026 mV/mm | -7.4 mm/°C | 1° | 0 | 2 mm | 84.3 mV |
| BDC-14 | 5/15/75 | 0.026 mV/mm | -7.4 mm/°C | 1 | 0 | 2 mm | |
| BDC-15 | 6/4/75 | 0.018 mV/mm | -9.3 mm/°C | 3° | 3° | 2.6 mm | 72.3 mV |
| BDC-16 | 6/4/75 | 0.028 mV/mm | -5.6 mm/°C | 2° | 2° | 1.7 mm | 95.5 mV |
| BDC-17 | 6/4/75 | 0.02 mV/mm | -7.6 mm/°C | 1° | 1° | 1.7 mm | 68.1 mV |
| BDC-18 | 6/5/75 | 0.019 mV/mm | -9.3 mm/°C | 57° | 26° | 13 mm | 69.9 mV |
| BDC-19 | 6/5/75 | 0.027 mV/mm | -8.2 mm/°C | 1° | 1° | 0.93 mm | 99.0 mV |
| BDC-20 | 6/5/75 | 0.025 mV/mm | -6.8 mm/°C | 2° | 2° | 2 mm | 68.4 |
| BDC-21 | 6/5/75 | 0.018 mmV/mm | -11 mm/°C | 1° | 1° | 2.9 mm | 154.8 mV |
| BDC-22 | 6/6/75 | 0.03 mV/mm | -4.6 mm/°C | 3° | 2° | 1.7 mm | 80.5 mV |
| BDC-23 | 6/6/75 | 0.021 mV/mm | -5.0 mm/°C | 3° | 3° | 3.5 mm | 63.4 mV |
| BDC-24 | 6/26/75 | 0.019 mV/mm | -8.5 mm/°C | 22° | 25° | 24.8 mm | 58.0 mV |
| Average value: | | 0.0223 mV/mm | 6.5 mm/°C | | | 5.87 mm | |

sensor group BB. The results from this group of sensors show significant improvement over previous designs. The hysteresis for this group is zero in almost all cases. Because of the inversion of the diaphragms, the temperature coefficient is reversed; however, it is also much smaller (2.12 mmHg/°C). Due to improved processing the imbalance in these transducers has also been improved, averaging -27.3 mV.

The sensitivity of this group is somewhat lower, 0.0091 mV/mmHg, due to a variation in etching. Based on the results of Tables 5, 6, 7, and 8 it may be seen that there is no correlation between sensitivity and temperature coefficient. The relationship between temperature coefficient and temperature hysteresis is somewhat less definite. Graphical analysis of the data for this group of transducers suggests that hysteresis is a random occurrence, possibly dependent on small misalignments of the diaphragm halves, or crystal defects. In general, transducers with temperature hysteresis have large temperature coefficient, however the reverse is not true.

The gold metallization on the CD transducers caused problems in many of the devices. Transients in the bridge voltage, caused by capacitive or inductive coupling, tended in some cases to puncture the oxide under metallization and cause partial shorts to the bridge. Because of this effect a new set of transducers was made, identical to these, except without the gold shield.

This new group of transducers was designated DD (Table 9). Due to a slightly longer etching time, the sensitivity of these devices was increased to an average of 0.0119 mV/mm Hg. The temperature coefficient is 1.94 mmHg/°C and the average

TABLE 8

**Sensors Designated CD**

| No. | Date | Sensitivity | Temp coeff | Hysteresis Temp | Hysteresis Pressure | Offset |
|---|---|---|---|---|---|---|
| CD-1 | 8/14/75 | Diaphragm | -0.0163 mV/°C | 0 | 0 | -15.8 mV |
| CD-1 | 8/18/75 | 0.009 mV/mm | 5.1 mm/°C | 2.7 mm | 0 | -62.6 mV |
| CD-2 | 8/14/75 | Diaphragm | -0.0163 mV/°C | | | -15.8 mV |
| CD-3 | 8/13/75 | Diaphragm | -0.0125 mV/°C | | | -14.8 mV |
| CD-3 | 9/15/75 | 0.0083 mV/mm | 0.52 mm/°C | 0 | 0 | -21.7 mV |
| CD-4 | 8/14/75 | Diaphragm | 0.015 mV/°C | | | -15.0 mV |
| CD-4 | 9/25/75 | 0.009 mV/mm | 1.59 mm/°C | 0 | 0 | -23.7 mV |
| D-5 | 8/14/75 | Diaphragm | 0.0087 mV/°C | | | -8.7 mV |
| CD-5 | 9/26/75 | 0.0075 mV/mm | 1.72 mm/°C | 0 | 0 | -17.3 mV |
| CD-6 | 8/14/75 | Diaphragm | -0.0087 mV/°C | 0 | 0 | -30.5 |
| CD-7 | 10/1/75 | 0.0105 mV/mm | 2.72 mm/°C | 0 | 0 | -22.0 mV |
| CD-8 | 10/22/75 | 0.0115 mV/mm | 2.59 mm/°C | 0 | 1 mm | -18.7 mV |
| CD-9 | 10/1/75 | 0.0095 mV/mm | 1.36 mm/°C | 0 | 0 | -25.3 mV |
| CD-10 | 10/13/75 | 0.007 mV/mm | -0.34 mm/°C | 0 | 0 | -23 mV |
| CD-11 | 10/13/75 | 0.0098 mV/mm | 5.69 mm/°C | 15.4 mm | 7.6 mm | -41.5 mV |
| CD-12 | | | | | | Noise |
| CD-13 | | | | | | Noise |
| CD-14 | 10/14/75 | 0.01 mV/mm | 1.55 mm/°C | 0 | 0 | -27.1 mV |
| CD-15 | 11/5/75 | 0.008 mV/mm | 2.66 mm/°C | 3 mm | 0 | -18.5 mV |
| CD-16 | 11/5/75 | 0.0095 mV/mm | 1.91 mm/°C | 2.6 mm | 0 | -17.4 mV |
| CD-17 | 11/5/75 | 0.011 mV/mm | 1.28 mm/°C | 0 | 0 | -26.0 mV |
| CD-18 | 11/5/75 | 0.007 mV/mm | 2.16 mm/°C | 0 | 0 | -35.7 mV |
| CD-19 | 11/5/75 | 0.0085 mV/mm | 2.76 mm/°C | 2.9 mm | 0 | -19.2 mV |
| CD-20 | 11/10/75 | 0.009 mV/mm | 1.17 mm/°C | 0 | 0 | -17.3 mV |
| CD-21 | 11/14/75 | 0.0073 mV/mm | 1.53 mm/°C | 1.7 mm | 0 | -15.9 mV |
| CD-22 | 11/17/75 | 0.0078 mV/mm | 1 mm/°C | | 0 | -15.5 mV |
| CD-23 | 11/19/75 | 0.0118 mV/mm | 2.12 mm/°C | 2 mm | 0 | -23.4 mV |
| CD-24 | 11/19/75 | 0.0085 mV/mm | 0.5 mm/°C | | | -17.0 |
| CD-25 | 11/19/75 | 0.010 mV/mm | 0.66 mm/°C | 0 | 0 | -17.3 |
| CD-26 | 11/26/75 | 0.0078 mV/mm | 2.5 mm/°C | 1.6 mm | 0 | -9.2 |
| CD-27 | 11/28/75 | 0.0115 mV/mm | -0.97 mm/°C | 2.2 mm | 0 | -21.2 |
| CD-28 | 11/28/75 | 0.0088 mV/mm | 1.4 mm/°C | 0 | 0 | -20.0 |
| CD-29 | 11/28/75 | 0.0103 mV/mm | 0 | 0 | 0 | -26.1 |
| CD-30 | 11/29/75 | 0.009 mV/mm | 1.05 mm/°C | 1.39 mm | 0 | -16.0 |
| Average values: | | 0.00915 mv | 1.80 mm/°C | | | 23.0 mV |

offset is −12.0 mV. Only 30% of the devices tested have measurable hysteresis. When packaged and coupled with the compensating amplifiers designed for these transducers, temperature coefficients should be reduced to less than 0.1 mmHg/°C; thus providing a useable transducer system.

The transducer amplifier has been designed as an interface circuit to provide stable amplification for the pressure transducer. Two basic circuit types have been developed – namely, a low-power, implantable circuit and a battery-powered amplifier for use outside the body. The implantable circuit design will be considered first.

Because of its small size, the transducer presents a few problems in actual measurement conditions. These problems include relatively large power consumption when powered by a steady DC current, low-level output requiring stable amplification, and large temperature coefficient of the output. The primary purpose of the implantable circuit, then, is to compensate for the transducer's electrical and thermal instabilities and to provide a stable higher level output for the telemetry applications. The design was carried out so that it could be fabricated eventually as a hybrid integrated circuit on a ceramic substrate using available integrated circuit chips.

A block diagram of the implantable interface

TABLE 9

**Sensor Designated DD**

| No. | Date | Sensitivity | Temp coeff | Hysteresis Temp | Hysteresis Pressure | Offset |
|---|---|---|---|---|---|---|
| DD-1 | 11/25/75 | 0.0108 mV/mm | 0.87 mm/°C | 1.2 mm | 0 | −3.7 mV |
| DD-2 | 11/26/75 | 0.0115 mV/mm | 0.20 mm/°C | 2.2 m | 0 | −6.2 mV |
| DD-3 | 11/26/75 | 0.013 mV/mm | 1.7 mm/°C | 0 | 0 | −13.8 mV |
| DD-4 | 11/26/75 | 0.0115 mV/mm | −0.94 mm/°C | 0 | 0 | −8.5 mV |
| DD-5 | 11/26/75 | 0.012 mV/mm | 0.07 mm/°C | 0 | 0 | −15.1 mV |
| DD-6 | 12/2/75 | 0.0143 mV/mm | 2.16 mm/°C | 0 | 0 | −5.0 mV |
| DD-7 | 12/3/75 | 0.0143 mV/mm | 1.9 mm/°C | 0 | 0 | −12.3 mV |
| DD-8 | 12/3/75 | 0.0095 mV/mm | 4.47 mm/°C | 0 | 2.6 mm | −15.3 mV |
| D-9 | 12/3/75 | 0.0115 mV/mm | 4.7 mm/°C | 19.5 mm | 2 mm | −20.3 mV |
| DD-10 | 12/3/75 | 0.0110 mV/mm | 2.42 mm/°C | 0 | 1 mm | −19.9 mV |
| DD-11 | 12/3/75 | 0.0115 mV/mm | 1.24 mm/°C | 0 | 0 | −3.9 mV |
| DD-12 | 12/4/75 | 0.0163 mV/mm | 2.46 mm/°C | 0 | 0 | −19.3 mV |

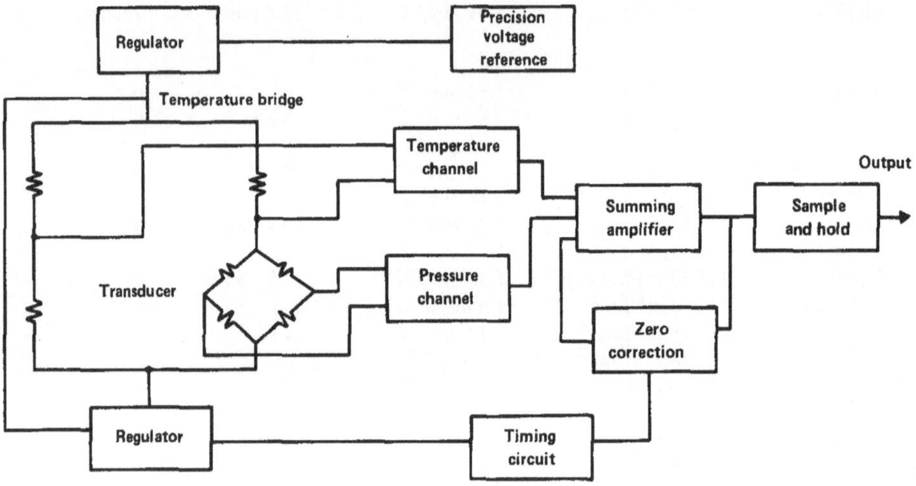


FIGURE 11. Interface circuit block diagram.

circuit design is shown in Figure 11. The circuit provides a pulsed, regulated voltage drive to the temperature bridge/transducer combination. The temperature signal and the pressure signal are independently amplified. A temperature-corrected pressure signal appears at the output of the summing amplifier. Zero correction is provided to prevent amplifier base-line drift. A timing circuit generates all the proper pulses to power the bridge, close the base-line correction loop, and enable the sample and hold module.

The method selected to temperature-compensate the transducer output is a double-bridge active compensation technique. The transducer is used as one leg of another outer bridge. The outer bridge is a temperature-sensing bridge. The four resistors of the pressure transducer form the inner bridge. In this circuit the thermal characteristic of the transducer bridge is employed to derive a temperature-dependent signal.

The major power consumer in the transducer/interface system is the transducer itself. At a voltage drive of 3.2 V, the current drain is approximately 1.2 mA. It would be very desirable to drop the current drain of the entire transducer/interface system to below 100 $\mu$A. The only approach to reducing the current consumed by the bridge is to pulse the bridge with low duty-cycle power rather than DC. It is impractical to drop the drive voltage considerably since the pressure output would be extremely small. To afford a really significant power consumption savings, the duty cycle of the bridge power pulses must be kept low.

The other consideration in power consumption is the current drawn by the amplifiers. The LM4250 micropower amplifiers can be biased at currents as low as 10 μA. However, at this current level, the gain bandwidth product is only about 50 kHz. The risetime of the pressure pulses is long because of low bandwidth. The bridge power pulses must be lengthened to accommodate the amplifiers. There is no reason, however, to keep the amplifiers idling during the time the bridge is not powered (more than 95% of the time).

The micropower amplifiers can be turned on and off quickly by pulsing the setting current resistor. The amplifiers need only be turned on when the bridge is powered. Because the amplifiers draw current with a low duty cycle, the pulse current can be quite high while maintaining low average current. By pulsing the amplifiers at high current with a low duty cycle, their full gain bandwidth can be realized (approximately 400 kHz). This increased bandwidth allows the bridge power pulses to be shortened. Thus, by pulse powering the amplifiers and the bridge, a significant decrease in power consumption can be achieved while actually increasing the system bandwidth.

To maintain constant transducer sensitivity with varying supply voltage, the drive on the transducer/temperature bridge combination must be held constant. Plus and minus power supplies are used to simplify the design of the base-line correction loop. Both the positive and negative drives on the bridge are regulated to power it effectively.

Both regulators require a temperature-stable voltage reference. The circuit chosen has a stability of better than 1% over a temperature range of 25 to 50°C.

As mentioned before, the temperature and pressure signals are independently amplified. In the summing amplifier stage, the temperature and pressure signals are combined in the proper ratio to produce a compensated output. Additionally, a base-line correction voltage is added to the summing amplifier to correct for input offset drift in the temperature channel, the pressure channel, and the summing amplifier itself.

A schematic drawing of the summing amplifier and base-line correction loop is shown in Figure 12. The summing amplifier is a standard inverting amplifier configuration. The more interesting part of this circuit is the base-line correction loop. The loop operates as follows. Upon a signal from the timing circuit, the inputs to both the temperature channel and pressure channel are switched to ground. Simultaneously, the analog switch is closed, forming a closed loop from the summing amplifier output to its input. The output of the summing amplifier is connected to the noninverting input of the zero correcting amplifier. The

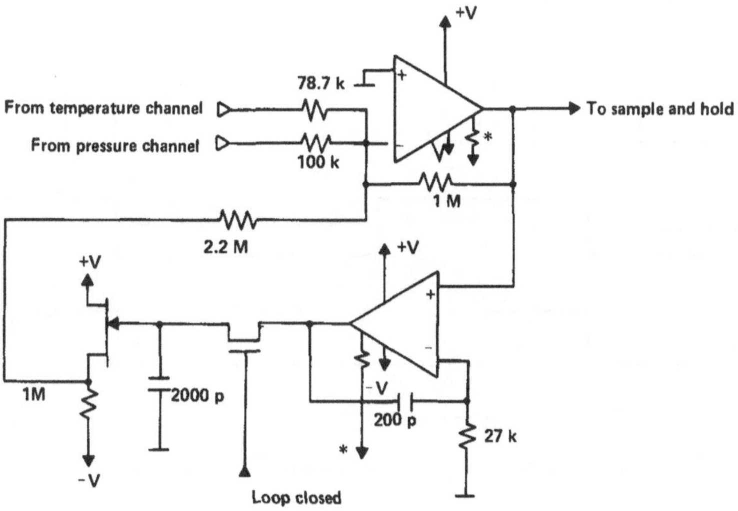


*To pulsed setting current.

FIGURE 12. Summing amplifier and base-line correction.

zero correcting amplifier drives the capacitor and the gate of the FET. The output of the FET, connected as a source follower, is summed with the temperature and pressure signals at the inverting input of the summing amplifier. Due to the extremely high gain of the zero correcting amplifier, the correction voltage at the summing amplifier changes until the output of the summing amplifier goes to zero (actually to the input offset voltage of the base-line correcting amplifier). After the circuit output is base-line corrected, the analog switch is opened and the timing circuit causes power to be applied to the double bridge circuit. (Refer to the timing diagram in Figure 13.) The correction voltage is held during the time the bridge is powered. In the present circuit, the bridge is powered for a 250-μsec period.

Because the bridge is only pulse powered, pressure information occurs in amplitude-modulated pulses. It is desired that the interface circuit provide a DC output corresponding to the pressure signal. Therefore, a sample and hold circuit was added to the interface circuit. The circuit consists of an analog switch, a holding capacitor, and a micropower buffer amplifier.

The timing circuit must provide pulses in the proper sequence to connect the bridge inputs to ground and close the correction loop, to open the correction loop and power the bridge, and finally to switch on the sample and hold circuit.

To allow for settling time in the interface circuit amplifier channels when the bridge is powered, a pulse length of 250 μsec has been chosen. The correction loop is closed for 250 μsec, then the bridge is powered and the correction loop is put into hold for the next 250 μsec. During the last 125 μsec of the bridge-powering interval, the sample and hold circuit is switched on. To accomplish this the amplifiers must be powered for 500 μsec.

The circuit chosen to accomplish this task consists of a CMOS astable multivibrator which clocks a CMOS counter. The frequency of the astable is 4 kHz, yielding a clock period of 250 μsec. By decoding the status of a seven-stage ripple counter with simple gating, all the proper timing signals are obtained.

A complete schematic diagram of the interface circuit is shown in Figures 14A and 14B. It is designed to operate from a nominal battery voltage of ±2.7 V. It will operate reliably down to a ±2.2-V supply. The number of capacitors has been kept to a minimum. Four are required for operation. This number will allow easier integration into hybrid circuit form, since chip capacitors are large.

The interface circuit itself has been constructed on a 3-½ in. X 4-½ in. printed circuit board for test and evaluation. A typical transducer mounted in a flat-pack was connected to the interface circuit inputs and put through a series of controlled-limit temperature cycles to determine the temperature coefficient of the compensated output. The measured output from the interface circuit is presented in Table 10. The sensitivity change with temperature is due to the change of transducer

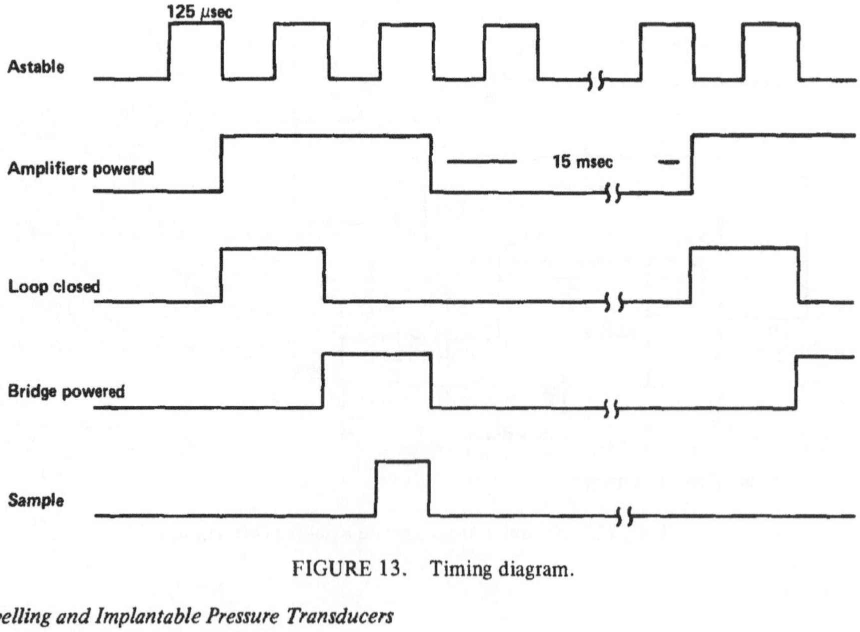


FIGURE 13. Timing diagram.

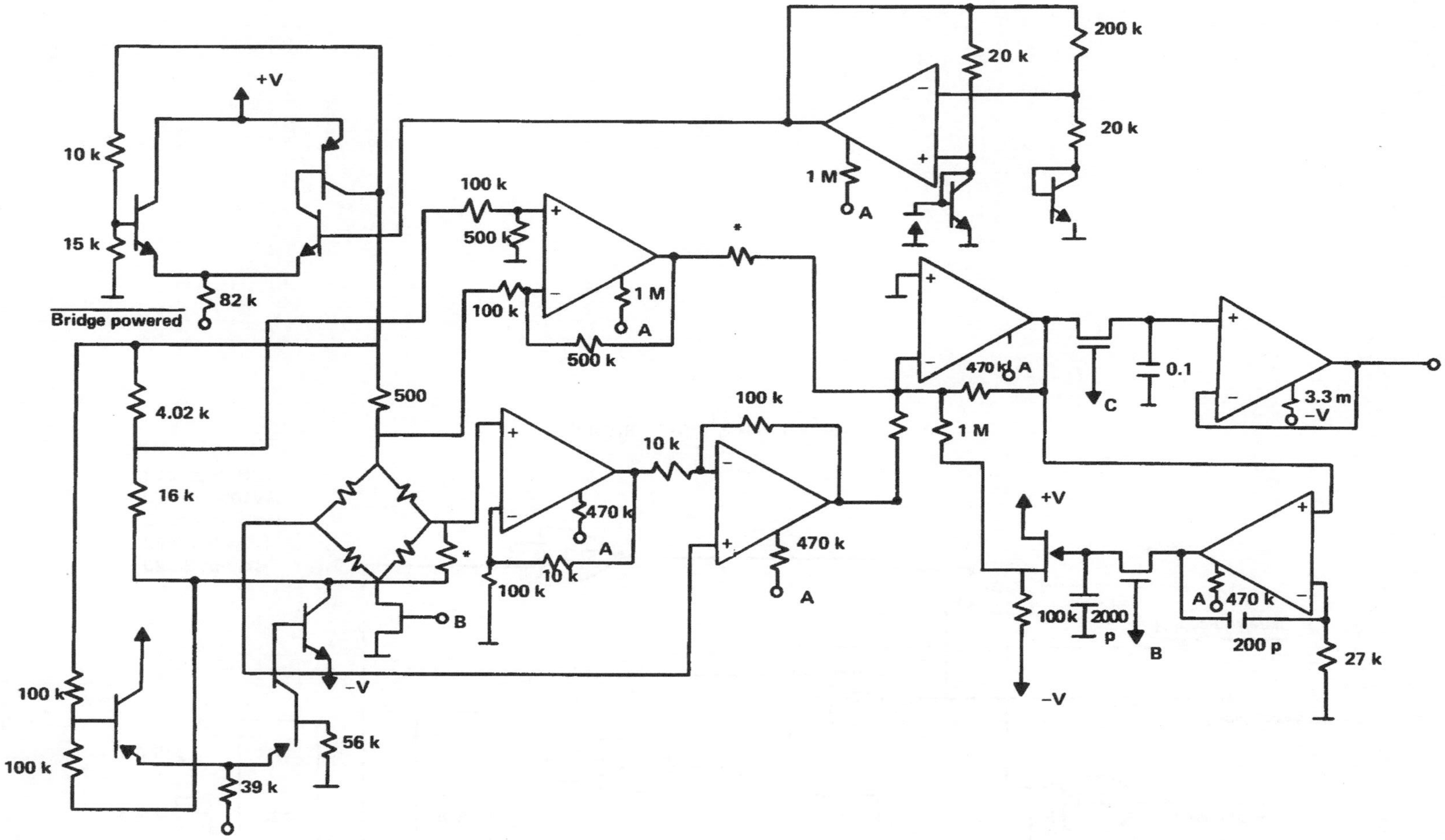


*Selected.

FIGURE 14A. System schematic. A. Amplifiers powered. B. Loop closed. C. Sample.

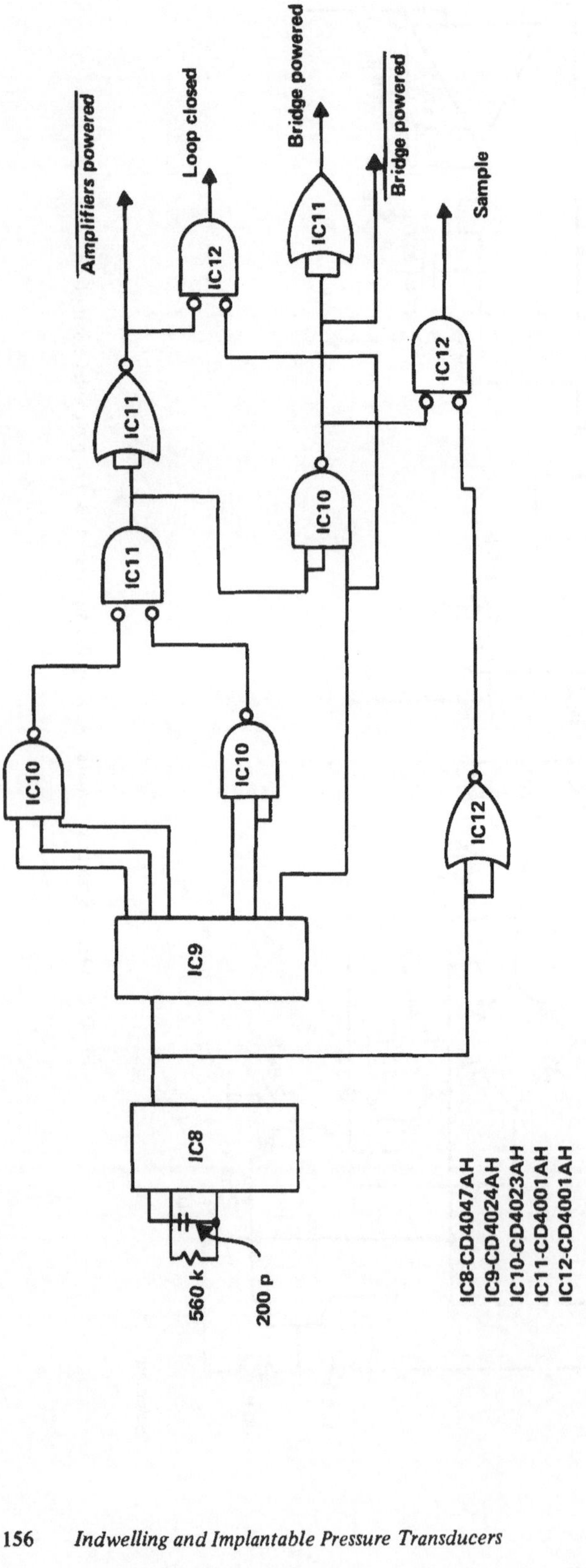


FIGURE 14B. System schematic.

TABLE 10

**Interface Circuit Performance Measurements**

| Parameter | Measured value |
|---|---|
| Pressure sensitivity ($S = \frac{\Delta V}{\Delta P}$) | 4.31 mV/mmHg |
| Output temperature coefficient ($\frac{\Delta P}{\Delta V}$) | ±0.1 mmHg/°C |
| Change of pressure sensitivity ($\frac{\Delta S/S}{T}$) | -0.167%/°C |
| Current drain | 60 μA |
| Linearity | Within 1% of full scale in 0- to 300-mmHg range. |
| Pressure sampling rate | 67/sec |
| Noise and uncertainty in output | 2 mV |
| Minimum supply voltage required | ± 2.2 V |

sensitivity with temperature and not due to varying drive voltages.

The interface circuit has been miniaturized and laid out on a 1 in. X 1 in. ceramic substrate. Commercial hybrid circuit components are being used in construction. The construction is underway and will soon be completed and evaluated.

A laboratory-type amplifier to evaluate transducers mounted to catheters was also constructed. The unit was designed to amplify and temperature-compensate all the various configurations of the pressure transducers developed at the Engineering Design Center. It was designed to offset correct transducers with initial offsets of 200 mV at 5.2 V drive. At maximum setting, the unit will temperature-compensate a transducer with 10 mmHg/°C temperature coefficient and 40 μV/mmHg sensitivity at $V_{drive} = 5.2$ V.

The design utilizes the same method of temperature compensation as the implantable circuit, the only difference being that the temperature bridge balance is adjustable to accomodate transducers of different total resistance. The amplifiers used in the circuit are low-noise, low-drift operational amplifiers; therefore, the need for base-line correction has been eliminated.

Temperature compensation adjustment is accomplished by thumbwheel switches mounted on the front panel. A simple calculation determines the number to be dialed into the switches for proper compensation.

A schematic drawing of the unit is presented in Figure 15. The circuit is designed to operate from a single 12-V rechargeable battery. The bridge is regulated to provide stable operation with varying battery voltage. A battery-status indicator circuit is included to give the user a quick indication of the condition of the battery.

The last part of the pressure transducer system is the package that holds the transducer for in vivo use. In order to be of practical use, this package must be as small as possible. In most cases the transducer will be introduced into the environment to be monitored via a catheter or small tube. Even in the case of telemetry the sensor will usually be separate from any other devices. Connections to the transducer are therefore made through four fine wires, which are welded to the bonding pads on the sensor. Experimentation showed that the best choice for these wires, considering strength and size, was 3-mil nickel wire. Teflon® was chosen as the wire insulation because of the requirement to keep the wire size small and maintain isolation. In order to weld the wires to the transducers the nickel is partially removed chemically and then gold-plated. The gold-plated wires are then spot-welded to the gold contact pads using parallel gap electrodes.

The primary requirement for the package must therefore be to insulate the wire attachments to the transducer.

The package must also provide mechanical protection for the transducer, as well as assuring biocompatibility. The simplest packaging possible is just an insulating coating over the wire-bonding area. Such a package requires only a material that is impermeable to body fluids and adheres to the transducer and wires to provide a seal. A variety of materials were evaluated as possible sealing compounds; among them were polyurethane, polyethylene, Teflon, epoxy, and silicone rubber. Parallel testing of both basic materials and coated

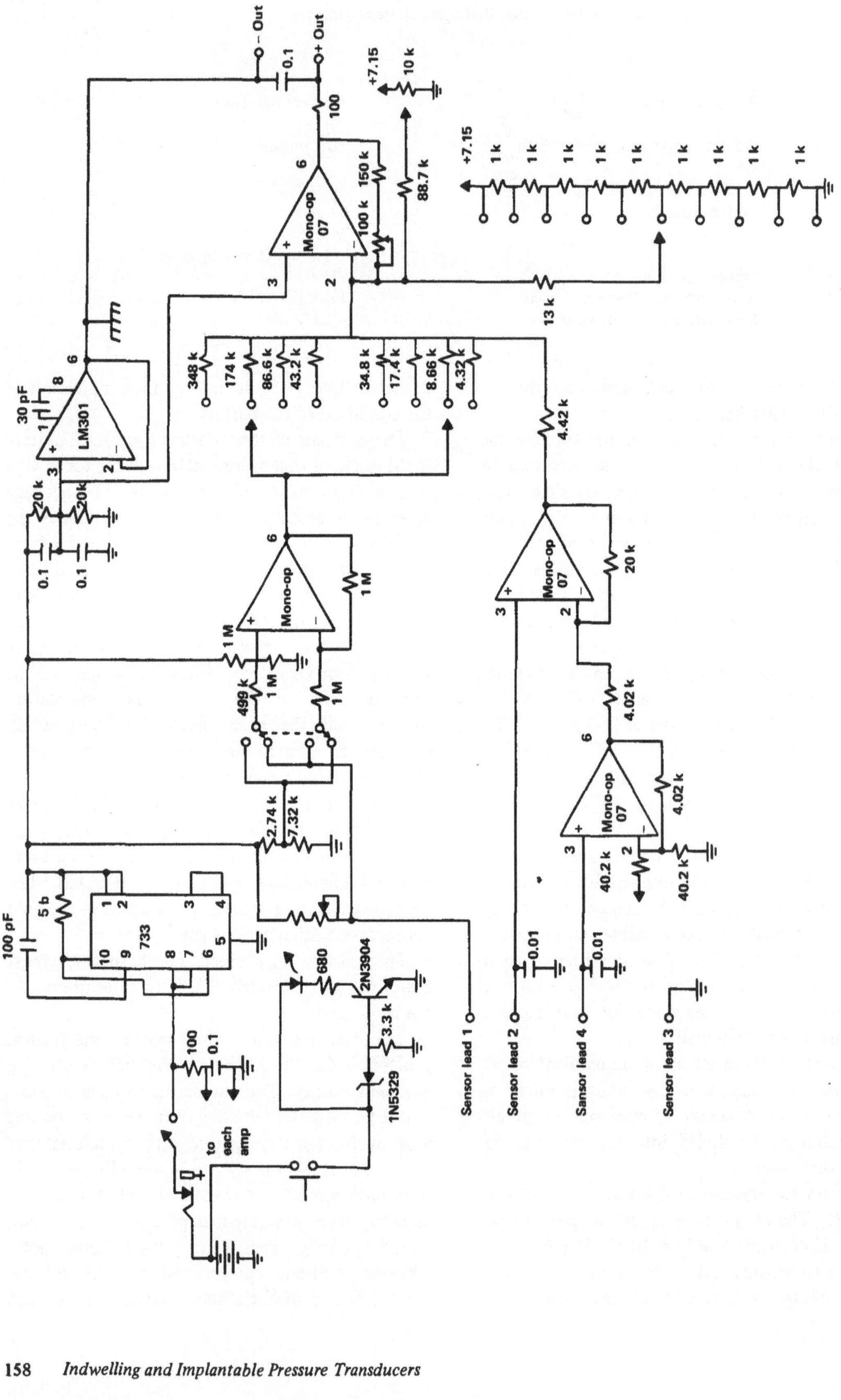


FIGURE 15. Temperature-compensating sensor amplifier.

transducers revealed that a useful package had several requirements beyond those just mentioned. A separate but coordinated effort has been made to locate and evaluate materials for a practical package (see section on "Material for Transducer and Implant Package").

In view of these tests conducted on packaging (Table 11), a set of base requirements has been formulated. Coating the sensors only at the wire attachment area was found unsuitable for two reasons. One, the coating would have to adhere with a fluid-proof seal to the silicon surface, including sharp edges and the gold-tin alloy seal. The resistance required between leads for less than 1 mmHg equivalent shift in output is 125 MΩ. Secondly, the package must give mechanical protection to the transducer, which is brittle. In view of this, it appears that the best package is one which completely encapsulates the transducer. In this manner there are no edges or seals to leak, and the transducer may be supported mechanically on one side. For this type of package the coating material must possess special characteristics. The coating must be soft enough to flex with the diaphragm, yet strong enough to hold the sensor and wires to their support. The elasticity of the material must be great enough to allow it to expand thermally along with the transducer without appreciably changing its thermal coefficient or hysteresis. Finally the material must be useable in thin coats, and it must still adhere to all of the packaging components.

The package which best serves all of these requirements is a silicone-rubber coating with a

TABLE 11

**Packaged Pressure Sensors**

| No. | Date | Sensitivity | Temp coeff | Hysteresis | Offset | Material |
|---|---|---|---|---|---|---|
| CNC-10 | 3/18/75 | 0.022 mV/mm | -6.4 mm/°C | 16.3 mm | 46.2 | G.Y. polyurethane |
| CNC-20 | 3/17/75 | 0.047 mV/mm | -4.3 mm/°C | 8 mm | 50.4 | G.Y. polyurethane |
| CNC-23 | 3/17/75 | 0.028 mV/mm | -6.3 mm/°C | 2.2 mm | 51.6 | G.Y. polyurethane |
| CNC-25 | 3/26/75 | 0.04 mV/mm | -6.8 mm/°C | 10.6 mm | 26.9 | G.Y. polyurethane |
| CNC-29 | 4/15/75 | 0.038 mV/mm | -4.8 mm/°C | 1.3 mm | 42.7 | G.Y. polyurethane |
| CNC-30 | 4/14/75 | 0.034 mV/mm | -4.4 mm/°C | 6.9 mm | 95.4 | G.Y. polyurethane |
| CNC-32 | 5/9/75 | 0.056 mV/mm | -3 mm/°C | 3 mm | 60.6 | G.Y. polyurethane |
| CNC-36 | 5/9/75 | 0.017 mV/mm | -5.6 mm/°C | 3 mm | 49.9 | 3140 RTV |
| BDC-1 | 5/19/75 | 0.02 mV/mm | -7.5 mm/°C | 28 mm | 65.2 | G.Y. & poly tube |
| BDC-2 | 5/27/75 | 0.025 mV/mm | -3.8 mm/°C | 7.1 mm | 60.3 | RTV & poly tube |
| BDC-4 | 8/18/75 | 0.017 mV/mm | -9 mm/°C | 8.9 mm | 85.3 | RTV & poly tube |
| BDC-9 | 9/15/75 | 0.002 mV/mm | 200 mm/°C | 312 mm | 74.7 | Parylene – 4 mil |
| BDC-10 | 9/17/75 | 0.006 mV/mm | Variable | >100 m | 37.5 | Parylene – 4 mil |
| BDC-12 | 9/18/75 | 0.007 mV/mm | 35.4 mm/°C | >200 mm | 37.7 | Parylene – 4 mil |
| BDC-15 | 9/26/75 | 0.006 mV/mm | 88.7 mm/°C | 133 mm | 65.0 | Parylene – 4 mil |
| CD-4 | 10/20/75 | 0.01 mV/mm | -2.63 mm/°C | 7.9 mm | 14.5 | RTV & Silastic® tube |
| CD-5 | 10/20/75 | 0.008 mV/mm | 0.14 mm/°C | 3 mm | 17.7 | RTV & Silastic tube |
| #28 | 10/16/75 | Diaphragm | -0.017 mV/°C | 0.05 mV | -11.7 | 50% Vorite® & DB |
| #28 | 10/18/75 | Diaphragm | 0.22 mV/°C | 12 mV | -12.2 | Vorite & poly tube |
| #29 | 10/16/75 | Diaphragm | 0.053 mV/°C | 0.075 mV | -15.6 | 50% Vorite & DB |
| #29 | 10/18/75 | Diaphragm | 0.33 mV/°C | 15 mV | -18.2 | Vorite & poly tube |
| #30 | 10/16/75 | Diaphragm | 0.033 mV/°C | 0.025 mV | -16.3 | RTV & Silastic tube |
| #32 | 10/17/75 | Diaphragm | 0.057 mV/°C | 0.025 mV | -14.6 | RTV & poly tube |
| #33 | 10/17/75 | Diaphragm | 0.017 mV/°C | | -16.2 | None |
| Y-1 | 10/8/75 | Diaphragm | 0.33 mV/°C | 12 mV | 39.8 | Parylene – 4 mil |
| CD-15 | 11/24/75 | 0.0083 mV/mm | 16.7 mm/°C | 15 mm | -19.4 | RTV & ceramic |
| CD-17 | 11/24/75 | 0.010 mV/mm | -3.58 mm/°C | 32.5 mm | -22.8 | RTV & ceramic |
| CD-18 | 11/25/75 | 0.0073 mV/mm | 3.0 mm/°C | 6.9 mm | -35.9 | RTV & ceramic |
| CD#22 | 11/25/75 | 0.008 mV/mm | -1.4 mm/°C | 6.3 mm | -18.5 | RTV & ceramic |
| CD#20 | 12/1/75 | 0.0085 mV/mm | -1.2 mm/°C | 4.4 mm | -17.6 | RTV & ceramic |
| DD#1 | 12/2/75 | 0.0118 mV/mm | 1.06 mm/°C | 3.2 mm | -2.8 | RTV & ceramic |
| DD#2 | 12/1/75 | 0.0123 mV/mm | -1.49 mm/°C | 6 mm | -8.5 | RTV & ceramic |
| DD#4 | 12/1/75 | 0.013 mV/mm | -1.7 mm/°C | 7.7 mm | -19.1 | RTV & ceramic |
| DD#5 | 12/2/75 | 0.013 mV/mm | 2.38 mm/°C | 2.9 mm | -17.2 | RTV & ceramic |

ceramic wafer providing physical protection for the transducer. The package is made by bonding with silicone rubber the transducer and its Teflon-insulated wires to an aluminum chip 50 mil wide by 300 mil long. The transducer and ceramic support are then dip-coated with more silicone-rubber sealant, covering the entire transducer and several inches of the attached wires. In most cases a plastic tube of approximately 50 mil diameter is slid over the wires down to the ceramic chip, and it is bonded with silicone-rubber adhesive. The tube prevents fluids from leaking along the Teflon to silicone interface, and it also provides mechanical protection for the attachment wires. Laboratory tests suggest that such a package should last at least 4 weeks in actual use.

At the present time a few transducers, packaged as described above, are being used experimentally and others are being prepared for additional testing. It is hoped that the data obtained from these tests will help to improve the present package. Also work is continuing on producing transducers with greater sensitivity and minimal hysteresis and temperature drift. The goal at this time is to produce a finished system that will have an unamplified sensitivity of 40 $\mu$V/mmHg, a compensated temperature drift of less than 0.05 mmHg/°C, and a base-line stability of 1 mmHg/day. Such a system would also be able to survive a minimum of 90 days in actual use.

## CONCLUSION

Development and construction of a miniature pressure transducer for biomedical application has been given. It includes design of the monolithic integrated circuit, as well as design of the signal processing circuitry.

A number of the testing procedures was also found to test the device parameters vital for the successful operation of the pressure transducer. It was possible to measure very small hysteresis effects in both temperature and pressure cycle and correlate these results with changes of the device structure.

The most difficult area of development was identified to be the final packaging. A complete and totally satisfactory solution to the problem of protecting the device against the body fluid environment remains yet to be solved.

Finally, Figures 16 to 23 were included to illustrate the various stages of the sensor fabrication procedure.

## ACKNOWLEDGMENT

This work was conducted under NIH Grant 6M14267. The author would like to thank the staff of the Engineering Design Center, particularly Scott Boettcher and John Homa, who greatly contributed to the development of the sensor in all its phases.

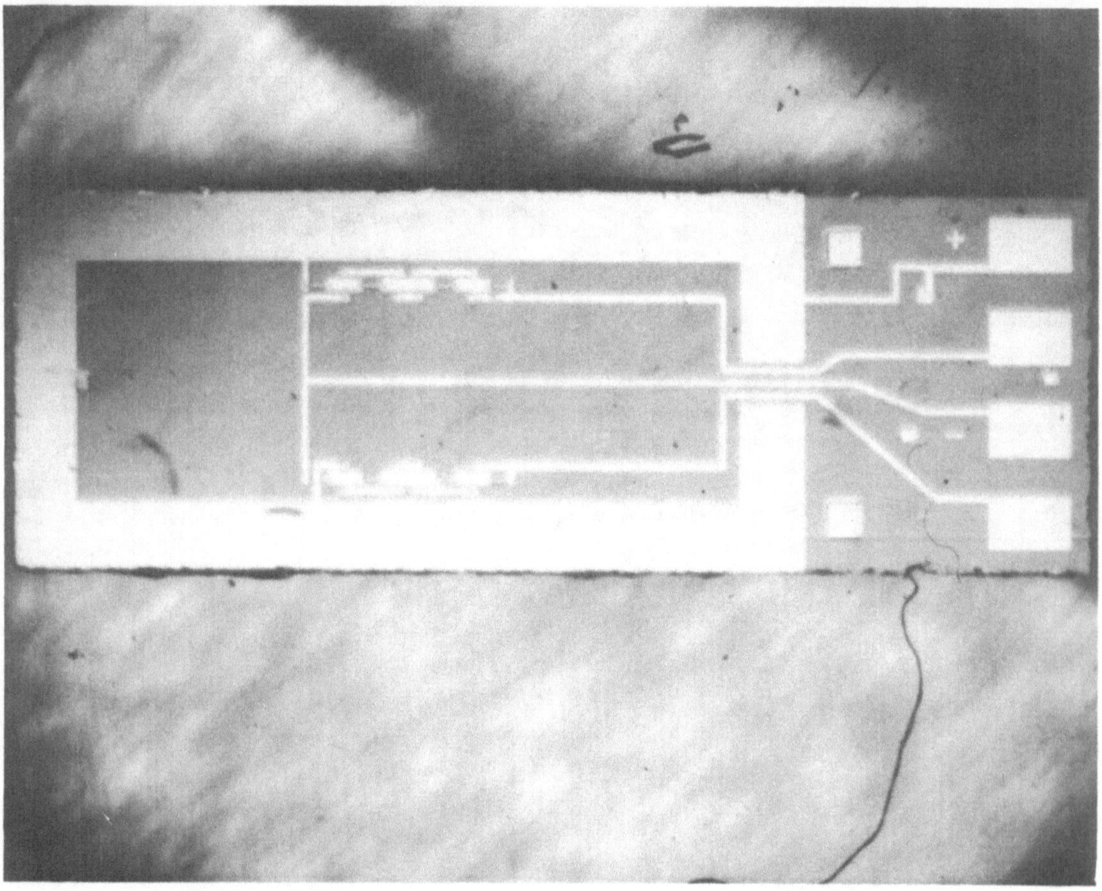

FIGURE 16. Sensor with resistors located at the edge of the diaphragm.

FIGURE 17. Sensor with resistors located at the center of the diaphragm.

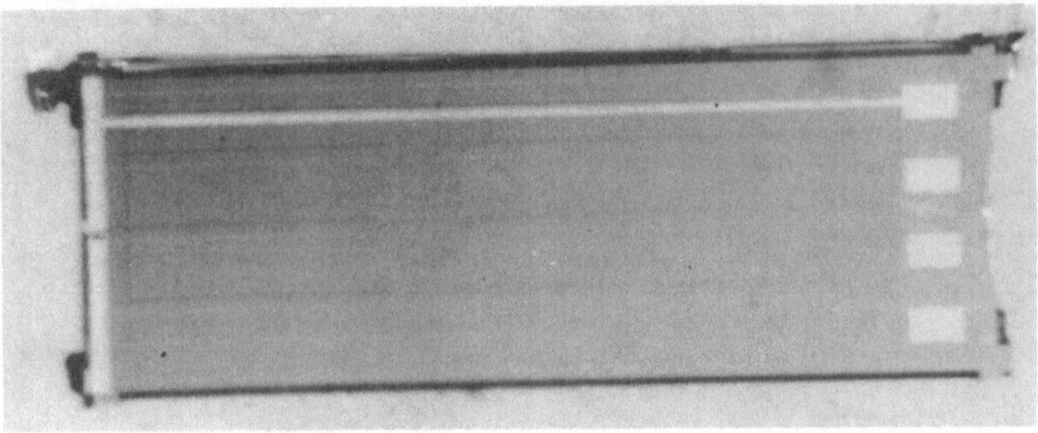

FIGURE 18. Sensor with resistors in the center and diffused interconnecting leads.

FIGURE 19. Sensor prior to vacuum sealing. Tin-gold preform is in the center of the figure.

FIGURE 20. Vacuum-sealed sensor with attached wires.

FIGURE 21. Protective overcoat applied to the finished sensor.

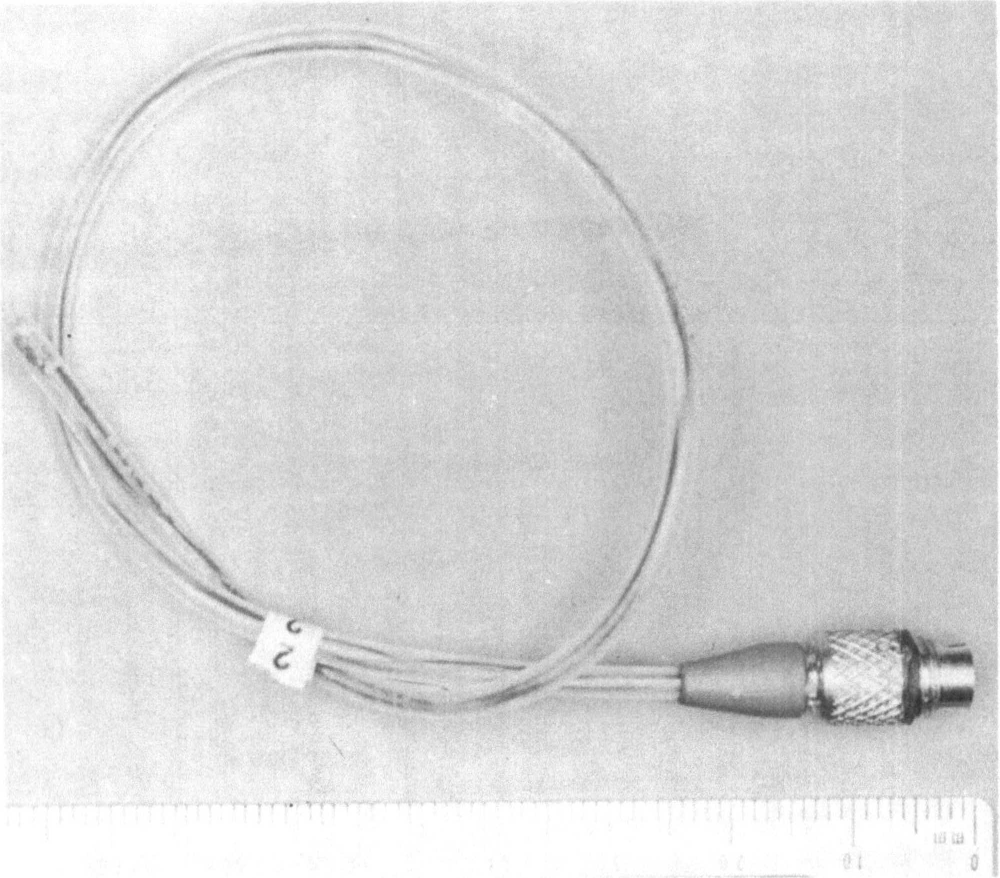

FIGURE 22. Sensor packaged in a catheter.

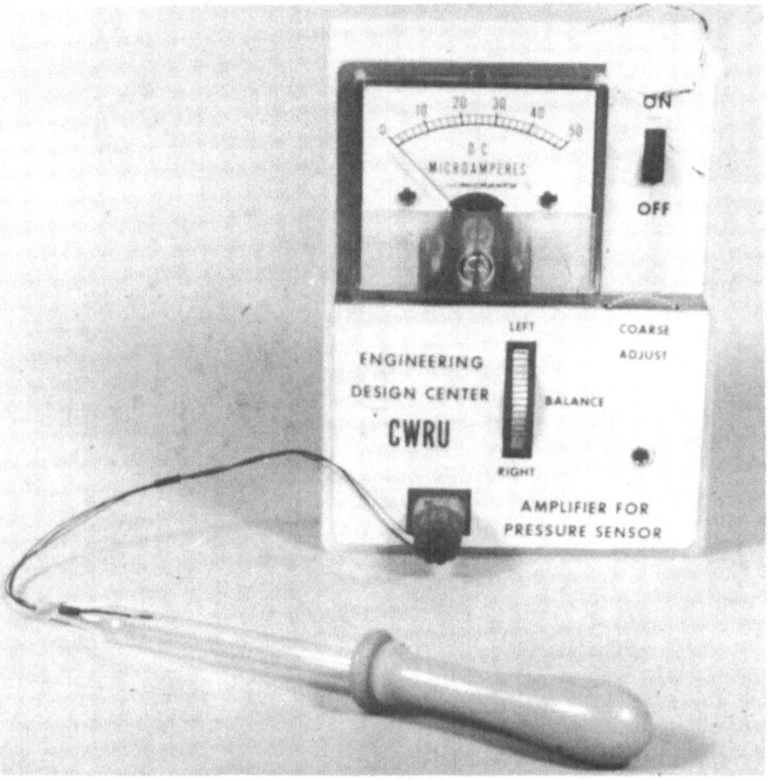


FIGURE 23. Signal processing amplifier attached to the sensor as a pressure-sensing demonstration unit.

# REFERENCES


1. Millar Mikro-Tip® Pressure Transducer, Millar Instruments, Inc., Houston, Texas.
2. Sensotec Division of Scientific Advances, Inc., Columbus, Ohio.
3. Kulite Semiconductor Products, Inc., Ridgefield, New Jersey.
4. **Ko, W. H.**, Tissue Reaction to Implanted Semiconductor Crystal, MEL Lab Memo No. 824, 1968.
5. Research Triangle Institute, Integrated Silicon Device Technology, Vol. V, ASD-TDR-63-316, Research Triangle Institute, Durham, North Carolina, 1964.
6. **Samaun, Wise, K. D., and Angell, J. B.**, An IC Piezoresistive Pressure Sensor for Biomedical Instrumentation, Department of Electrical Engineering, Stanford University, California, 1972.
7. **Lee, D. B.**, Anisotropic etching of silicon, *J. Appl. Phys.*, 40, 4569, 1969.
8. **Taroni, A., Prudenziati, M., and Zanarini, G.**, Semiconductor sensors, *IEEE Trans. Ind. Electron. Control Instrum.*, IEC1-17, 415, 1970.
9. **Mason, W. P. and Thurston, R. N.**, Use of piezoresistive materials in the measurement of displacement, force, and torque, *J. Acoust. Soc. Am.*, 29, 1096, 1957.
10. **Tugte, O. N., Chapman, P. W., and Long, D.**, Silicon diffused-element piezoresistive diaphragms, *J. Appl. Phys.*, 33, 3322, 1962.
11. **Pfann, W. G. and Thurston, R. N.**, Semiconducting stress transducers utilizing the transverse and shear piezoresistance effects, *J. Appl. Phys.*, 32, 2008, 1961.
12. Available from Interman Industrial Products, 350 Old Country Road, Garden City, New York 11530.
13. **Schaffzin, R. A.**, Structural isolation of a constrained diaphragm, *Instrum. Control Syst.*, p. 88, April 1972.

# PIEZOELECTRIC FIELD-EFFECT TRANSDUCERS*

**Keming W. Yeh and R. S. Muller**

Strain transducers consisting of a piezoelectric medium coupled directly to an insulated-gate field-effect-transistor structure are described. This construction leads to high strain sensitivity and a very localized sensitive area (800 $\mu m^2$ in present devices). The transducer fabrication process is compatible with that used for silicon integrated circuits. This makes possible mass production at low cost and integration with other signal processing circuitry.

Several different structures in this transducer family are described and evaluated for adaptability to medical applications. Emphasis is placed on the piezoelectric diffused MOS (PI-DMOS) transducer. This transducer provides improved strain sensitivity at high frequencies. Fabrication procedures and design rules are described and experimental results are given.

## I. INTRODUCTION

Research has been active for several years at the University of California, Berkeley, on strain transducers which make use of direct coupling between a piezoelectric medium and the channel of a field-effect transistor. Three embodiments of these piezoelectric field-effect transducers (PI-FET) have been built and studied thus far. The first kind, developed by Muller and Conragan,[1] utilizes an oriented piezoelectric film in a thin-film-transistor structure. It has special potential for medical applications because it can be deposited on a flexible membrane in arrays that can be interrogated by a clocked digital source. The second type of transducer, developed by Conragan and Muller,[2] has a piezoelectric substrate. The third type, developed by Greeneich and Muller,[3] has a MOSFET structure with a piezoelectric layer as the gate insulator. This last type has exhibited the highest stress sensitivity, a gauge factor of 9,000, and has detected strains of $10^{-8}$ at a frequency of 5.6 MHz. For better performance, a new structure, the PI-DMOS transducer,[4] has shown an even higher gauge factor, on the order of 100,000, and has exhibited high speed capability with response time of less than 400 ns. The piezoelectric thin-film transducer will first be described in Section II. Section III deals with the design and the fabrication technology of the PI-DMOS transducer. The theory of its operation and its measured performance are discussed in Section IV. In the last section some consideration is given to the temperature stability of the PI-DMOS transducer.

## II. PIEZOELECTRIC THIN-FILM FIELD-EFFECT TRANSDUCER

The first embodiment of a piezoelectric field-effect transducer was in the form of a CdS thin-film transistor. A full description of the device has been published, and its application as a microphone[5] was demonstrated. These transducers were successfully fabricated on thin polymer film material and the frequency response of the devices was determined and theoretically explained.[6]

Thin-film FET transducers have special potential for applications in the biomedical field. What is lacking at present is a thorough control of the processing parameters for these structures – essentially the kind of development effort that can only be mounted successfully by an industrial firm. Recent concentration on parameter control and reliability for thin-film active circuits has been reported at the Westinghouse Research Laboratory.[7] The procedures employed there, namely computer control of all depositions and elimination of most hand-processing steps, would impact very favorably the production of thin-film FET transducers.

## III. DESIGN AND FABRICATION OF PI-DMOS TRANSDUCER[8]

### Device Structure

A cross-sectional view of an n-channel PI-DMOS transducer and a top view of the device are shown in Figure 1. The structure is similar to that of a D-MOS transistor[9] except that the gate $SiO_2$ is

*Research sponsored by National Science Foundation Grant ENG74-17688-A01.

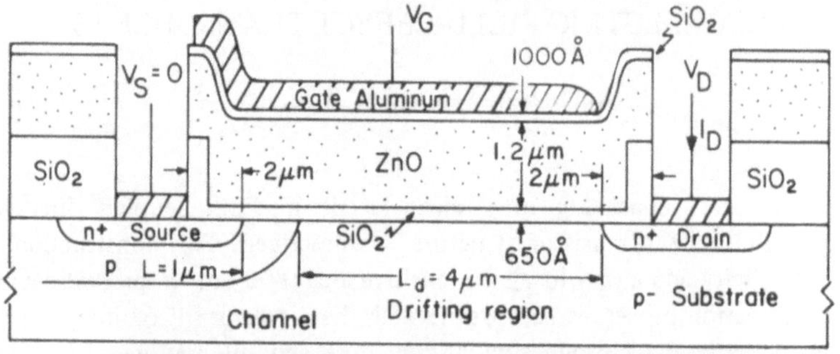


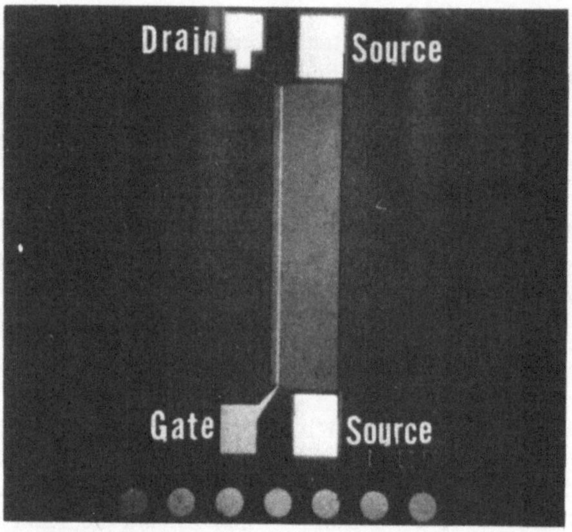


FIGURE 1. The cross-sectional view and the top view of an n-channel PI-DMOS transducer. (From Yeh, K. W., Fabrication technology of a piezoelectric double-diffused MOS (PI-DMOS) transistor, in *Proc. Int. Microelectronic Symp.*, International Society for Hybrid Microelectronics, Montgomery, 1975, 453. With permission.)

layered and overlain with a thin piezoelectric ZnO film that is sandwiched between the thermally grown $SiO_2$ layer and a sputtered (or vapor-deposited) $SiO_2$ layer. Zinc oxide was chosen for the piezoelectric because of the extensive research already reported about preparing oriented layers of the material. The substrate is a lightly doped p-type silicon having a resistivity of 15 Ω-cm. The lateral net impurity profile is very similar to that of a conventional diffused npn bipolar transistor. The p-type channel-region doping is graded. The oriented ZnO film is deposited by RF sputtering. The bottom thermally grown $SiO_2$ layer forms the coupling capacitance from the strain signal to the channel and protects the silicon surface during the sputtering process. The top $SiO_2$ layer inhibits any charge injection into the ZnO layer from the gate metal, acts as a mask for ZnO etching, and also passivates the ZnO layer.

**Design Rules**

Conclusions regarding optimization of device design as obtained from the theoretical analysis are listed below:

1. Sensitivity is increased by optimizing the electrical gain of the D-MOS transistor, which is accomplished by reducing channel length and thinning gate oxide layers.
2. Sensitivity is heightened by increasing the electromechanical constant of the piezoelectric medium.
3. Bandwidth is increased by reducing conductivity in the piezoelectric layer and by reducing the output capacitance of the device.
4. Speed of response is enhanced by shortening device channel length and reducing output capacitance.
5. Current capability is enhanced by a short

drift region, a large channel doping gradient, and an n-type drift region.

**Device Fabrication Technology**

The starting material is a 15 Ω-cm, p-type silicon wafer with (100) orientation. The sequential steps used in fabricating an n-channel PI-DMOS transducer are listed below:

1. Grow field oxide.
2. Open field-isolation region and diffuse $p^+$ impurity in oxidizing atmosphere.
3. Open source window and diffuse p-type channel impurity in nonoxidizing environment.
4. Open drain diffusion window and diffuse $n^+$-type impurity over source and drain regions in oxidizing atmosphere.
5. Open gate region and grow thin layer of HCl silicon dioxide.
6. Open source and drain contact windows.
7. Deposit ZnO piezoelectric film by RF sputtering process.
8. Deposit $SiO_2$ layer by RF sputtering process and anneal the wafer.
9. Etch top $SiO_2$ and ZnO layer over the source and drain contact windows.
10. Aluminum metalization and sintering.

The technological aspects of optimizing the design parameters are discussed in the following:

**ZnO film deposition** – This is the most crucial step, since the strain sensitivity of the device and the threshold voltage shift due to the RF sputtering process are determined at this point. Good control of optimized sputtering parameters must be maintained to obtain oriented polycrystalline ZnO piezoelectric films.[8,10]

**Channel length control** – The length of the drift region is mainly determined by the original mask. For a well-controlled channel length, the source window edges must be undisturbed during the channel impurity diffusion. If this is done in a $N_2$ atmosphere at high temperature, silicon-nitride pits usually appear at the silicon surface. To avoid pits, a slight $O_2$ flow is maintained during channel diffusion. The thin oxide formed is removed by an HF dip before the source impurity diffusion. The channel length thus obtained is found to be reproducible.

**Threshold voltage control** – The device threshold voltage is determined by the peak channel doping concentration which is dependent on the doping levels in the substrate, the channel, and the source, and on the indiffusion profiles. For n-channel devices, it was found that p-type substrates provide for a better control of threshold voltage.

**Channel doping gradient** – It is desirable to have a high channel doping gradient. For an n-channel device, this is accomplished by shortening the channel length rather than by increasing the peak channel doping concentration, because a "boron skin" forms under heavy boron doping.

**$SiO_2$ layer thickness** – Thin $SiO_2$ layers are desirable. The bottom $SiO_2$ should be thick enough to protect the silicon surface during RF sputtering, yet still thin enough to provide a large charge-coupling capacitance. A thickness of 600 Å is used. The concern at the thin top $SiO_2$ layer is pinholes which may result in charge injection from gate metal into ZnO layer.

## IV. DEVICE OPERATION AND PERFORMANCE

When a stress is applied externally, it induces a piezoelectric polarization potential in the ZnO layer. The polarization charge is coupled to the inversion channel in the FET through the $SiO_2$ capacitance. The carrier concentration in the channel and the drain current are thus modulated by the piezoelectric polarization. The stress signal is detected and amplified by the associated D-MOS structure. The response time is mainly determined by the frequency performance of the D-MOS, which is limited by the transit time and the output capacitance. (The present device has a limiting frequency in the gigahertz range.) The delay in the drift region has little effect since carriers travel through it at saturated velocity under normal device operation.[9] For a drift-region length of 4 μm this delay is 61.5 psec, assuming a saturated carrier velocity of $6.5 \times 10^6$ cm/sec.

The I-V characteristic of the device is governed by the following first-order nonlinear differential equations:[11]

$$I = [f(v) + \delta g(x, V_p)] \frac{dv}{dx} \qquad (1)$$

where f(v) represents the conductivity due to the electrical bias applied to the device, $\delta g(x,V_p)$ represents the contribution to conductivity due to piezoelectric polarization, and $V_p$ is the piezoelectrically induced polarization gate voltage. The

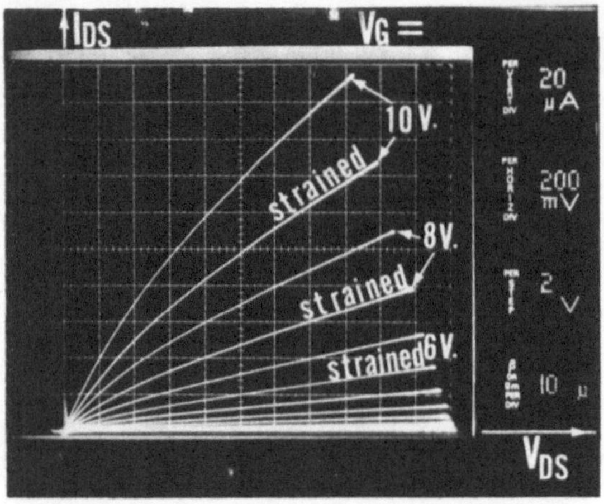


Strain (s) = $1.8 \times 10^{-6}$

Gauge Factor $\dfrac{\Delta I_D/I_{DO}}{S} = 1.6 \times 10^5 \ (V_{GS} = 8V)$

FIGURE 2. PI-DMOS response to a constant stress. (From Yeh, K. W., Fabrication technology of a piezoelectric double-diffused MOS (PI-DMOS) transistor, in *Proc. Int. Microelectronic Symp.*, International Society for Hybrid Microelectronics, Montgomery, 1975, 453. With permission.)

fractional change in saturation drain current as a function of piezoelectric polarization induced by constant stress is given by

$$\frac{\Delta I_D}{I_{DO}} = \frac{2V_p}{(V_G - V_T)} = \frac{2S \cdot a}{(V_G - V_T)} \qquad (2)$$

when $V_p$ is much smaller than $V_G - V_T$, as is usually the case. In Equation 2 $V_T$ is the threshold voltage, S is strain, and "a" is a combination of the elastic, piezoelectric, and dielectric constants of the ZnO medium. A measure of the strain sensitivity of the pressure sensor, gauge factor, can be derived from Equation 2 for the PI-DMOS transducer.

$$G.F. \equiv \frac{\Delta I_D/I_{DO}}{S} = \frac{2a}{V_G - V_T} \qquad (3)$$

The strain sensitivity of the device is demonstrated in Figure 2, where a constant stress is applied by bending the free end of a cantilever beam on which the device is bonded by silver epoxy. The gauge factor calculated there is much larger than that available from commercial sensors.

When an AC stress is applied to the device, the fractional change in drain current when the device is operated in the linear region is given by

$$\frac{\Delta I_D}{I_{DO}} = \frac{V_{pO}}{(V_G - V_T) - \frac{V_D}{2}} \frac{\sin(kL/2)}{(kL/2)} e^{-i\omega t} \qquad (4)$$

where k is the wave factor of the AC stress signal and L the channel length of the device. Figure 3 shows device response to an AC stress at 600 Hz which was applied through a vibrator to the above-mentioned cantilever beam structure.

To demonstrate its speed capability, the device was mounted on the inside wall of a shock-wave cylinder and was exposed to a shock wave with an equivalent strain of $10^{-6}$ at the wavefront. The corresponding response through drain output voltage is shown in Figure 4. The fluctuations were due to the improper mounting of the device and were later eliminated by fixing the device edges with nonconductive epoxy. A rise time of 400 ns was clearly observed, which was due to the large parastic capacitance in the nonoptimized measuring circuitry. With an improved circuitry the rise time should be no more than 20 ns.

The device can also be used as a detector for either bulk acoustic waves transmitted through the silicon substrate, or for surface acoustic waves launched at the surface of the silicon chip. For bulk acoustic wave detection, it has been established that the device has constant gain over a wide frequency range.[11] The detection characteristics for surface acoustic waves are being examined at present.

## V. DEVICE STABILITY

Solid state pressure sensors have generally exhibited high strain sensitivity and a large instability problem resulting from the environment-sensitive nature of the material physical constants. In the category of passive devices, a bridge structure is usually used to compensate for the zero drift and the temperature coefficient, while preserving the strain sensitivity. For active devices, such as MOS transistors, proper choice of the DC electrical operating condition can be used to obtain zero temperature coefficient, and a bridge structure may also be used for additional stability. Due to the extra piezoelectric layer, further complication is involved for the PI-DMOS transducer.

In principle, the physical constants (elastic, piezoelectric, and dielectric) of the piezoelectric medium have negligible temperature coefficient. The main instability of PI-DMOS transducer then

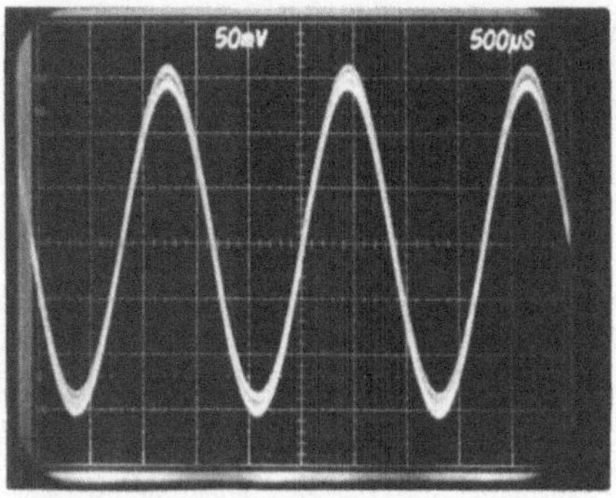


FIGURE 3. PI-DMOS response to AC stress at 600 Hz. $V_T = 4.1^V$; $V_{DD} = 10$ V; $V_{GG} = 13$ V; RL = 33 kΩ.

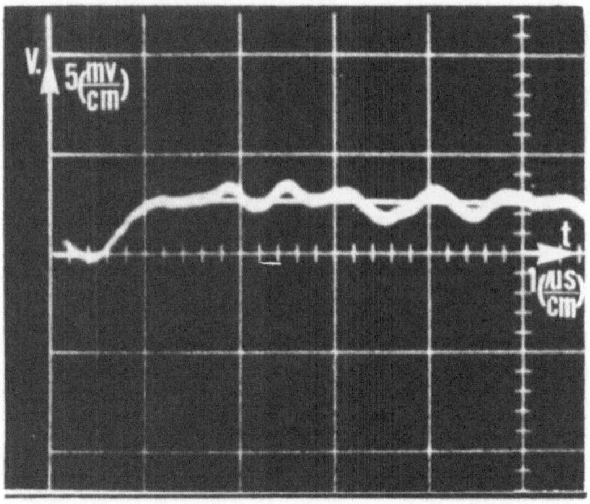


FIGURE 4. PI-DMOS response to a shock wave with an equivalent strain of $10^{-6}$ at the wavefront. (From Yeh, K. W., Fabrication technology of a piezoelectric double-diffused MOS (PI-DMOS) transistor, in *Proc. Int. Microelectronic Symp.,* International Society for Hybrid Microelectronics, Montgomery, 1975, 453. With permission.)

comes from the associated D-MOS structure in the substrate silicon which can be minimized by the proper choice of electrical operating condition. In practice, the long-term and short-term instabilities mainly come from the ZnO piezoelectric film. When the ZnO film and the top $SiO_2$ film are deposited by RF sputtering, radiation damage is introduced at the silicon surface due to the energetic particles in the sputtering process. It causes zero drift and can be effectively removed through proper heat treatment[8] (part of the radiation damage can be removed by annealing at 450°C for 10 min in $N_2$ or $O_2$ environment). Contaminants in the sputtering chamber may be incorporated into film during the sputtering process and cause puzzling instability phenomena.
incorporated into film during the sputtering pro-
Oxygen, hydrogen ions, and water vapor are easily adsorbed on the ZnO surface and cause both zero drift and a degradation of the piezoelectric response. A passivating layer after the metalization process is necessary to prevent this adsorption.

A preliminary result has shown that the strain sensitivity of the device is constant over a temperature range from −68° to 150°C when the above preventative steps are taken. No noticeable zero drift has been observed over a 3-month period. More extensive studies are necessary before a conclusion can be reached.

## VI. CONCLUSIONS

Different types of field-effect transducers have been described and compared. They exhibit a high strain sensitivity within a small active area. With the incorporation of the D-MOS transistor structure, both strain sensitivity and response speed are greatly improved. The PI-DMOS transducer has shown a gauge factor of $10^5$, which is 20 times better than a PI-FET and about 1,000 times better than commercial gauges. The response time under optimized operating condition is expected to be less than 20 ns.

The design rules and fabrication technology of PI-DMOS devices have been discussed. The stability problem has been investigated with some encouraging preliminary results. Although more research is required, it is felt that this family of pressure transducers is more attractive for external precision pressure monitoring than for implantable applications.

# REFERENCES


1. **Muller, R. S. and Conragan, J.**, A metal-insulator-piezoelectric semi-conductor electromechanical transducer, *IEEE Trans. Electron Devices*, ED-12, 590, 1965; U.S. Patent 3,351, 786, November 7, 1967.
2. **Conragan, J. and Muller, R. S.**, Piezoelectric field-effect strain transducers, in *IEEE Proc. Solid-State*, Catalog No. 70C25, SENSOR, Institute of Electrical and Electronics Engineers, New York, 1970, 52; U.S. Patent 3,585,415, June 15, 1971.
3. **Greeneich, E. W. and Muller, R. S.**, Acoustic wave detection via a piezoelectric field-effect transducer, *Appl. Phys. Lett.*, 20, 156, 1972.
4. **Yeh, K. W. and Muller, R. S.**, Piezoelectric Diffused MOS (PI-DMOS) Transistor as Strain Transducer, 1975 Device Res. Conf., Ottawa, Canada, June 24–26, 1975.
5. **Muller, R. S., Conragan, J., and Fiebiger, J.**, MIPS transducer microphones, *Proc. IEEE*, 54, 337, 1966.
6. **Fiebiger, J. R. and Muller, R. S.**, Electrical performance of metal-insulated piezoelectric semiconductor transducers, *J. Appl. Phys.*, 38, 1948, 1967.
7. **Cresswell, M., Sienkiewicz, L. J., Bogel, G. F., Yu, K. K., Csakvary, T., and Rogers, W. L.**, An Integrated All Thin-Film Digital Timer, Paper WAM 3.5, presented at Int. Conf. on Solid-State Circuits, Philadelphia, February 18–20, 1976.
8. **Yeh, K. W.**, Fabrication technology of a piezoelectric double-diffused MOS (PI-DMOS) transistor, *Proc. Int. Microelectronic Symp.*, Orlando, Florida, International Society for Hybrid Microelectronics, Montgomery, 1975, 453.
9. **Sigg, H. J., Vendelin, G. D., Cauger, T. P., and Kocsis, J.**, D-MOS transistor for microwave application, *IEEE Trans. Electron Devices*, ED-19, 45, 1972.
10. **Yeh, K. W., Muller, R. S., Wu, W-K, and Washburn, J.**, Characterization of RF sputtered ZnO piezoelectric films using transmission electron microscope, in *Proc. of the Thirty-Third Annual Meeting of the Electron Microscopy Society of America*, Las Vegas, Nevada, Claitor's Publishing Division, Baton Rouge, 1975, 86.
11. **Yeh, K. W.**, Piezoelectric Diffused MOS (PI-DMOS) Strain Transducer as Surface Acoustic Wave Detector, Ph.D. thesis, University of California, Berkeley, 1976.

# CAPACITANCE PRESSURE TRANSDUCERS

**Thomas B. Fryer**

The conversion of mechanical quantities such as pressure to electrical signals for remote readout and recording requires a suitable transducer. Changes in the circuit parameters of resistance, capacitance, and inductance have all been utilized by designers at various times in the past. Resistance strain gauge bridges are widely used because of their stability, but are not sensitive enough for miniature low-pressure biomedical applications. To obtain greater sensitivity, solid state silicon-diffused piezoresistive elements have been used, but with some sacrifice in long-term stability. The need for improved accuracy in low-pressure measurements makes it desirable to examine other possible transducer elements. But, for long-term stability, all the problems have not been solved. Capacitance and inductance transducers have been used in the past, but require more complex circuitry than the DC voltage used with a resistance bridge, and may offer some stability advantages. A mechanical design that is stable both with temperature and time (without creep) is of course essential, and the inert materials, metals, and ceramics needed for a capacity system make this technique attractive. For low pressures and small devices, the few picofarads that are obtained and the small percentage variation that results present a difficult challenge for the electrical measurement. Shielding of distributed capacitances is essential, and the availability of integrated circuits that can be built into the transducer helps eliminate shielded cable problems. The use of innovative circuits such as described by Harrison[1] or as part of the LC tank circuit of an oscillator offers great potential when coupled with a suitably designed pressure transducer.

The type of pressure transducer described here is one using a conventional (elastic diaphragm) conversion system, such as the one discussed in the chapter by L. A. Geddes, in which the mechanical motion is detected and converted into an electrical signal.

Now, the conversion to an electrical signal offers a number of possibilities, in both active and passive elements. Resistors, capacitors, and inductors are passive elements that can be utilized. Obviously, the greatest emphasis has been on the resistive elements because one can do this with thin film, thick film, and integrated circuit-type elements, and the resistance values are usually very stable. There are also active elements that offer possibilities, but they are subject to considerable drift. Even the solid state strain gauges diffused into diaphragms have been weak as far as stability problems are concerned. So, coming back to the passive elements, we must recognize that the capacity variation, although not always easy to work with because it is small in magnitude and distributed capacity effects, may be (with proper circuit designs) useful for biomedical transducers.

Now a description will be provided of two circuits that we have developed over a period of years that seem to work well with low-capacity type transducers. We did see one transducer (described in chapter by A. E. Walker, L. J. Viernstein, and J. G. Chubbuck) that was a capacity-type transducer. There Walker and co-workers used bellows, in other words, a mechanical leverage-type arrangement to obtain a larger capacity change. This simplifies the electrical measurement problem, but obviously there are disadvantages, especially size. The device that we have used is a capacity cell that is exactly similar to a Konigsberg 7-mm button type pressure transducer.

Unfortunately, a great deal of data are not available on many of the units used for the purposes of demonstrating the stability, but Figure 1 shows a basic circuit for a capacity transducer. The unit is excited with an oscillator, an AC signal, typically in the 100,000 kHz to 1 MHz range. This is a capacity bridge and its main feature is the use of a diode quad. The two diodes, $D_1$ and $D_2$, do not have to be in the same location as $D_3$ and $D_4$, because they are connected in such a way that the capacitive effect across the bridge does not affect

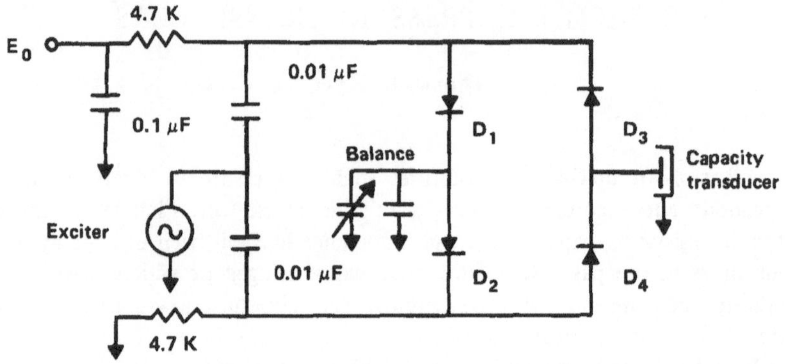


$$\frac{\Delta E_0}{E_p} = -\frac{\Delta C}{C_0}$$

**Transducer 81**
$C_0$ = 7.8 pF
(Exc. = ±10 V; f = 2 MHz)

| $E_0$ mV | P mmHg |
|---|---|
| −543 | 0 |
| −412 | 9.6 |
| −310 | 16.2 |
| −212 | 22.1 |
| −109 | 28.4 |
| −009 | 34.2 |
| +103 | 40 |
| +203 | 44.8 |
| +206 | 50.5 |
| +400 | 51 |
| +500 | 55.3 |

**Transducer 75**
$C_0$ = 7.0 pF
(Exc. = ±10 V; f = 2 MHz)

| $E_0$ mV | P mmHg |
|---|---|
| −564 | 0 |
| −411 | 4.7 |
| −306 | 7.5 |
| −199 | 10.1 |
| −100 | 12.6 |
| 0 | 15.1 |
| +100 | 17.5 |
| +202 | 19.7 |
| +300 | 21.8 |
| +401 | 23.8 |
| +510 | 26.0 |

Note: For 34 mmHg, $\Delta C \cong 0.39$ pF; for 15 mmHg, $\Delta C \cong 0.35$ pF.

FIGURE 1. Micro capacitor bridge circuit.

the output reading. So, the two diodes, $D_3$ and $D_4$, can be made through integrated circuits with either a thick-film or thin-film type of arrangement and placed right on the transducer. This is the essential feature that is now available to us that was not available when this circuit was first used almost 10 years ago. Because of the commercial availability of the strain gauge transducers, we had dropped the work with capacity cells for a number of years. As a result of all the problems with the stability, it appeared to be desirable to take another look at this circuit as a means of obtaining a stable device, especially since the transducer itself is so simple. The transducer consists of two plates and it can be essentially a monolithic structure with a glass-to-metal seal, which is what we use. This is a transducer which was constructed for Dr. Harold Sandler of the NASA Ames Research Center; it was built into the end of a regular catheter guidewire that is 0.0040 in. in diameter. In the case of the catheter we are not looking for high long-term stability, rather we are looking for size and resolution. Therefore, we have a trade off of size vs. stability. It might be surprising to note that the capacity device has only a 1-pF total capacity, the variation is probably 0.1 pF and we can measure down to $10^{-15}$ to $10^{-16}$ F. So, resolution is not a problem and the entire unit can be built right into the transducer because of solid state technology.

Now, the rest of the circuit in Figure 1. One can apply a signal (exciter) and have a balancing capacitor here, and the diodes rectify the signal so that the resulting DC bias changes at $E_0$ as a function of pressure or capacity change. So it is a standard capacity-type bridge, but by building in the solid state devices right at the unit many cable problems can be eliminated. The diaphragm on this unit uses about a 2- to 5-μm metal that is stretched and welded across the surface, and the spacing is on the order of half a thousand, or say, 10 to 12 μm. Now, this one type of circuit will give a DC voltage the same as a resistance bridge and requires an AC oscillator or exciter.

Figure 2 shows a diagram of the catheter, an intra-arterial pressure transducer. There is the diaphragm, with two diodes placed inside the housing. Two sides of the bridge wiring are brought out with two coaxial cables, and this is all located inside a guidewire with the sleeving to protect it.

Figure 3 demonstrates an intracranial pressure application, where we have long-term stability needs and a much larger size pressure cell is feasible. The pressure cell used is almost identical physically in size with the 6- to 7-mm strain gauge cuff type transducer that is commercially available. The capacity cell is used as a part of an oscillator circuit. However, the oscillator circuit is integral with the device so that there is no distributed capacity with the cable, and the advantages of this are quite considerable for this application.

The entire system requires only battery power to obtain the 10 MHz variable oscillation frequency. Besides, the tank circuit, which is of course the primary frequency-changing element, there are second-order effects due to the transistor and some loading effects that can cause small frequency changes. In order to eliminate these undesirable effects, the first stage is coupled to a second stage which then radiates out the frequency and isolates the loading effects from the actual tank circuit. The entire circuit is inside a sealed chamber or package. The LVA 450 Zener diode and two transistors are used to provide a regulated supply voltage.

We can also supply the system with an inductively coupled energy source instead of using a battery. The output is a frequency that is stable and unalterable once generated, and there is no bridge balancing to change from one unit to the next. The output is easily read on a frequency counter. Consequently, one is not always concerned with such problems as: did I have the same bridge, has it been changed, or where is the drift problem coming from?

Table 1 shows the calibration data over a period of a couple of weeks for units constructed in May 1975, and as can be seen, we were getting roughly a 7,600-Hz change in frequency per millimeter of mercury. Obviously there is no problem with sensitivity; it is stability or signal to noise that is always the problem.

Over a period of a couple of weeks the equivalent change was of the order of ½ mmHg. Figures ranging from ½ to 2 mmHg a month have been reported at this conference, but I don't think they are too meaningful, because stability is very unpredictable. Most changes are of a random nature and cannot be extrapolated. Changes may occur frequently one week and very little the next, except perhaps for some consistent drift problems. The transducer unit providing the data shown in

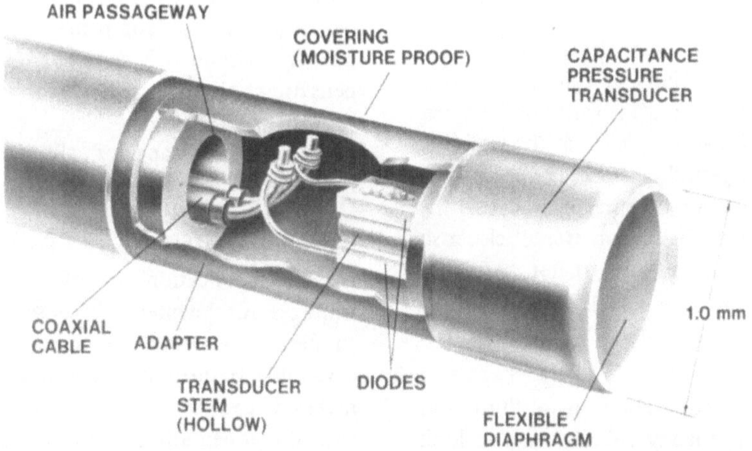


FIGURE 2. Intra-arterial pressure transducer.

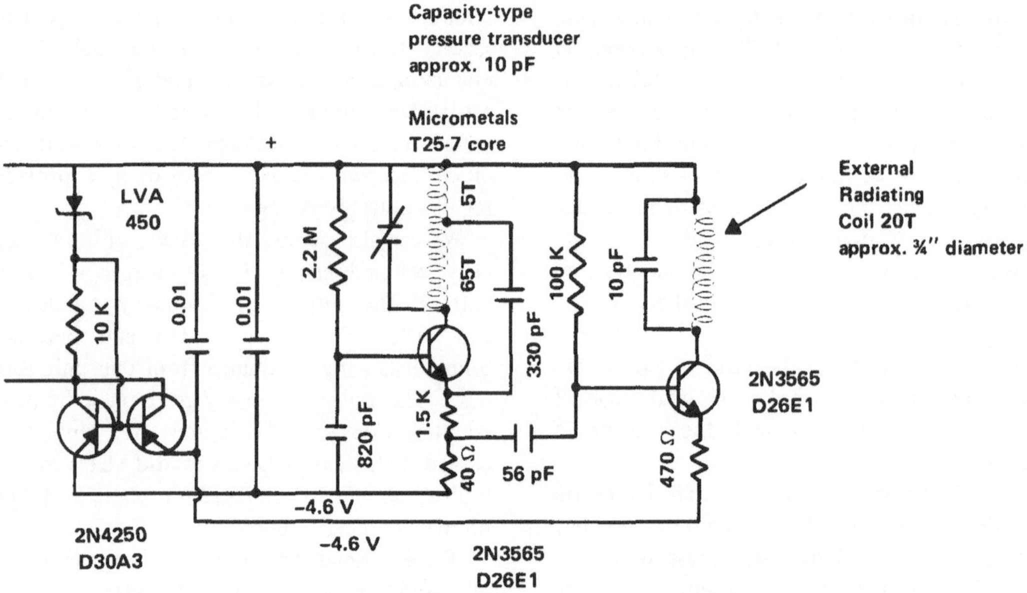


FIGURE 3. Pressure transducer system capacity-type frequency variable circuit.

Table 1 was not used for 6 months after its construction, but then it was implanted in an animal three or four times and taken out. Drs. A. K. Ream and G. O. Silverberg, of the Stanford Medical Center, said that the unit was mishandled and that no special care was taken with it. We do see a variation here that represents about 10 to 15 mm of change. So, we do not have the progressive data which we will obtain in the future.

The two transducers used were made 10 years ago, so they were not current ones. New units are being built so that we will have some more statistical data as far as the quantity of units, but the results look very encouraging.

The data from a second transducer are presented in Table 2. Units of transducers were made at the same time, and the variation here was considerably less. It can be seen that the variation here over a 2-week period is almost imperceptible. Now this has a thinner diaphragm, because the sensitivity is higher or the spacing is a bit closer so that the capacity changes are greater per millimeter. The sensitivity is about 18.4 kHz so changes on the order of ½ mm over a period of a couple of weeks were seen.

Table 3 presents the data from a third unit made up without a capacity cell, in order to look at the stability of the circuit. This has a standard 10-pF, good quality type of capacitor in it. The equivalent, in terms of pressure, to the kind of sensitivity that we have in the cells of Tables 1 and 2 would be of the order of 1 mm over 6 months. The circuit itself is perfectly capable of handling this type of capacity change or this type of electrical signal. Also, the transducers themselves are welded and are essentially monolithic. They are made with a Kovar glass-to-metal seal, and hopefully they can be evacuated and made into an absolute cell so that we do not have the temperature variations of the air pressure change. The temperature effects, which were not pointed out on the other two, are not more than ½ mm or less per degree. Hence, compensation is not necessary, but the circuit of course could have a compensating circuit built into it without any problems.

However the capacity cell, because it does not have any of the glues or the other problems that seem to be some of the major sources of drift, can be a very stable device. For the intracranial pressure application, it offers some unique advantages for telemetry because the circuit is so simple. The system is very simple since there is just a variable frequency oscillator that is easily telemetered and does not require any bridges or special compensation to convert the bridge circuit into a suitable telemetry signal.

TABLE 1

**Calibration Data for Variable Capacity Transducer with Frequency Variable Oscillator (#1 Unit with Pressure Cell)***

| Date (1975) | Temperature (°C) | Pressure (mmHg) | Oscillator (MHz) | Frequency change (kHz) |
|---|---|---|---|---|
| May 9 | | | | |
| Initial reading | 20 | 0.0 | 9.8636 | |
| After 1 min | | 0.0 | 9.8555 | |
| After 1 hr | | 0.0 | 9.8556 | |
| May 12 | | | | |
| Initial reading | 20 | 0.0 | 9.8648 | |
| After 2 hr | | 0.0 | 9.8603 | |
| | | 0.0 | 9.8594 | |
| | | | | 7,600 |
| | | 1.0 | 9.8518 | |
| | | | | 7,400 |
| | | 2.0 | 9.8444 | |
| | | | | 7,800 |
| | | 3.0 | 9.8366 | |
| | | | | 7,700 |
| | | 4.0 | 9.8295 | |
| | | | | 7,700 |
| | | 5.0 | 9.8218 | |
| | | | | 75,800 |
| | | 10.0 | 9.7836 | |
| | | | | 81,300 |
| | | 20.0 | 9.7023 | |
| | | | | 87,400 |
| | | 30.0 | 9.6149 | |
| | | | | 91,000 |
| | | 40.0 | 9.5239 | |
| | | | | 101,700 |
| | | 50.0 | 9.4222 | |
| | | 0.0 | 9.8566 | |
| | 30 | 0.0 | 9.8132 | |
| | 41 | 0.0 | 9.7575 | |
| May 13 | | | | |
| After 9:30 | 21 | 0.0 | 9.8545 | |
| After 1:00 | 20 | 0.0 | 9.8588 | |
| May 23 | 20 | 0.0 | 9.8586 | |
| November 26 | | 0.0 | 9.618 | |
| | | 1.0 | 9.611 | |
| | | 2.0 | 9.604 | |
| | | 3.0 | 9.597 | |
| | | 4.0 | 9.591 | |
| | | 5.0 | 9.583 | |
| | | 10.0 | 9.549 | |
| | | 20.0 | 9.472 | |
| | | 30.0 | 9.391 | |
| | | 40.0 | 9.299 | |
| | | 50.0 | 9.202 | |

*Pressure sensitivity ≅ 7.7 kHz/mmHg. Equivalent pressure measurement stability over 2 weeks was better than 1 mmHg.

TABLE 2

**Calibration Data for Variable Capacity Transducer with Frequency Variable Oscillator (#2 Unit with Pressure Cell)***

| Date (1975) | Temperature (°C) | Pressure (mmHg) | Oscillator (MHz) | Frequency change (kHz) |
|---|---|---|---|---|
| May 5 | 20 | 0.0 | 10.3153 | |
| May 6 | | | | |
| Initial reading | 20 | 0.0 | 10.3420 | |
| After 1 hr | | 0.0 | 10.3244 | |
| May 7 | | | | |
| Initial reading | 20 | 0.0 | 10.3404 | |
| After 1 hr | | 0.0 | 10.3267 | |
| May 9 | 20 | 0.0 | 10.3365 | |
| May 12 | | | | |
| After 1 hr | 20 | 0.0 | 10.3270 | |
| | | 0.0 | 10.3260 | |
| | | | | 18,400 |
| | | 1.0 | 10.3076 | |
| | | 2.0 | 10.2888 | |
| | | 3.0 | 10.2695 | |
| | | 4.0 | 10.2491 | |
| | | 5.0 | 10.2291 | |
| | | 10.0 | 10.1250 | |
| | | 20.0 | 9.8910 | |
| | | 30.0 | 9.6206 | |
| | | 40.0 | 9.2936 | |
| | | 50.0 | 8.8602 | |
| | | 0.0 | 10.3150 | |
| | 28 | 0.0 | 10.2897 | |
| | 35 | 0.0 | 10.2545 | |
| | 40 | 0.0 | 10.2243 | |
| May 13 | | | | |
| Initial reading | | 0.0 | 10.3220 | |
| After 30 min | | 0.0 | 10.3164 | |
| May 23 | | 0.0 | 10.3059 | |
| November 26 | | 0.0 | 10.107 | |
| | | 1.0 | 0.090 | |
| | | 2.0 | 0.073 | |
| | | 3.0 | 10.056 | |
| | | 4.0 | 10.038 | |
| | | 5.0 | 10.019 | |
| | | 10.0 | 9.927 | |
| | | 20.0 | 9.724 | |
| | | 30.0 | 9.481 | |
| | | 40.0 | 9.183 | |
| | | 50.0 | 8.823 | |

*Pressure sensitivity ≅ 18.4 kHz/mmHg. Equivalent pressure measurement stability was better than 1 mmHg over 2 weeks.

## DIALOGUE

**Question**: What is the current drain of the circuit operating on the transducer at 6 V?

**Dr. Fryer**: Well, I am operating the circuit at around 4 V. That is because I am inductively powering, therefore voltage is of no concern. We have run that kind of circuit at 1½ V, but it is now set to run on the order of a milliampere, say 5 to 6 mW. However, such circuits have been run as low as 100 or 50 μA. The stability might not be quite as good but in this case it is around a milliampere.

**Question**: What is the temperature sensitivity of the circuit which you used?

**Dr. Fryer**: The temperature effect on the first

TABLE 3

**Calibration Data for Variable Frequency Oscillator Fixed 10-pF Capacitor in Place of Capacity Pressure Transducer**

| Date (1975) | Temperature (°C) | Power supply (V) | Oscillator (kHz) | Frequency change (kHz) |
|---|---|---|---|---|
| April 30 | | 5.5 | 9675.5 | |
| | | | | 4.5 |
| | | 6.0 | 9679.0 | |
| | | | | 1.5 |
| | | 6.5 | 9680.5 | |
| | | | | 0.6 |
| | | 7.0 | 9681.1 | |
| | | | | 0.1 |
| | | 7.5 | 9681.2 | |
| | | | | 0.2 |
| | | 8.0 | 9681.4 | |
| | | | | 0.5 |
| | | 10.0 | 9681.9 | |
| | | | | 1.2 |
| | | 15.0 | 9683.1 | |
| | | 7.6 | 9679.5 | |
| | 22 | 7.6 | 9688.6 | |
| | 32 | | 9640.4 | |
| | 42 | | 9595.6 | |
| May 1 | 22 | 7.6 | 9700.0 | |
| May 2 | 21 | 7.9 | 9784.9 | |
| May 5 | 20 | 7.35 | 9664.5 | |
| May 6 | | | | |
| Initial reading | 20 | | 9666.6 | |
| After 1 hr | | | 9675.8 | |
| May 7 | | | | |
| Initial reading | 20 | 7.6 | 9667.5 | |
| After 1 hr | | 7.6 | 9674.1 | |
| May 8 | | | | |
| Initial reading | 20 | 7.6 | 9674.0 | |
| After 4 hr | | 7.6 | 9677.4 | |
| End of day | 21 | 7.6 | 9672.6 | |
| May 9 | | | | |
| Initial reading | 20.5 | 7.6 | 9672.2 | |
| After 4 hr | 20 | 7.6 | 9677.2 | |
| May 13 | | | | |
| Initial reading | 20 | 7.6 | 9670.5 | |
| After 4 hr | | 7.6 | 9673.5 | |
| May 14 | 20 | 7.6 | 9673.8 | |
| May 23 | | | | |
| Initial reading | 22.5 | 7.6 | 9664.8 | |
| After 1 hr | | 7.6 | 9668.0 | |
| November 22 | | 7.6 | 9655.8 | |

test units is very small as seen in Tables 1 and 2. These cells were vented so there was not a temperature effect due to gas expansion. If the cell was sealed at atmospheric pressure, such an effect would exist. We expect to vacuum seal the cells to eliminate this effect. The effect of temperature on the circuit is very small.

**Dr. Lorig**: I would like to know, with respect to your circuit and then the ultimate model (which I assume has not been put into animals), if you have made some provision for not having distributive capacitors associated with your transducer. When you go to install this into an animal or into a human, you will have parasitic capacitance caused by the tissues and the biological fluids.

**Dr. Fryer**: No, that is, you have to understand

that this is all ground plane, everything else is inside. So, that is the difference between what you saw Dr. Walker* describe, where you have distributed capacity from the outside and you have got to radiate out. To radiate energy out, a buffered amplifier is used, so there are not any distributed capacity effects.

**Dr. Stone**: I was wondering if the tests which you have done so far are static tests.

**Dr. Fryer**: Well they are static. We have used them in animals to look at intracranial pressure variations.

**Dr. Stone**: All right, well there are two questions. Number one: in this particular application you show drift, per se; where is this drift coming from? Number two: what does the data show now about long-term stability?

**Dr. Fryer**: The data where we had 10 mm or so over a long period of time, where this thing has been put in or taken out and used a number of times.

**Dr. Stone**: As I saw it was in the arterial catheter unit.

**Dr. Fryer**: No, this was a complete intracranial unit. The data on both of them was the intracranial unit, and the first one (arterial catheter) was with the other circuit (with the DC voltage), which is what you might use on a catheter. DC voltage is obtained, just like with a Konigsberg cell with the leads.

**Dr. Sandler**: Let me answer some of that. We found that when we were using the manometer-tip sidebar, the original transducer was made for taking a Kovar housing and overlapping the weld; it had to be a stretch diaphragm so that the technology was very delicate. Also, it had to be a very thin diaphragm – the thinner the diaphragm the more sensitive the device. After we used it a couple of times with the animals and the data which we recorded 8 years ago when we did the original work, we found that the saline solution was etching the diaphragm of the transducer and we were getting significant variations. So we had to move to gold-plating the diaphragm. Hence, there are two places where this particular transducer has a problem: (1) in the uniform construction of the transducers and the overlapping of the welding of the diaphragm to the Kovar housing, and (2) in the protection of the diaphragm from corrosion by the biological fluids. I think that the technology has some of the answers to these problems.

**Mr. Brosh (Kulite, Inc.)**: Are these answers to the small catheter stability?

**Dr. Sandler**: There is no question that a 1-mm transducer can be constructed in the laboratory, but can it be reproduced in quantity from a production standpoint? In the original uses reported in 1967 we were attempting to sense a 1/10 pF at the tip with a catheter that had a capacitance which was also variable every time the catheter was moving. Until we develop a system to use the diode quad circuit, this is nearly an impossible thing.

**Dr. Fryer**: A production engineer would be needed to convert this kind of technology into a design suitable for production. It is normal for most companies to take a research design and redesign it before starting a new item in production. I am sure that you are aware of that. Modifications are often needed to use new technology or fabrication techniques and to make it practical from a production and cost standpoint.

**Edward Cheng**: You mentioned putting your inductive coil inside the package; I assume it will be included inside the Kovar-to-glass sealed capsule. Can you describe the final size and dimensions of the package?

**Dr. Fryer**: The size will be a 14-mm bore hole and the dimension in the other direction would be roughly 1.5 cm. Now it could be larger in diameter, similar to the one which Dr. Walker* described, and less than 1.5 cm thick. The geometry of Dr. Walker's unit is approximately that of a nickel, but thicker. A nickel is pretty large, it is over 2 cm in diameter.

**Mr. Konigsberg**: We have had some experience in building catheters similar to the one described. It will come as no surprise to many people in the audience if it is said that the transducer is not an end in itself. Connecting it to a cable and using the proper lead wires to put the whole thing together so that it is a functional and a reasonably inexpensive unit to make is a problem which requires a considerable amount of attention. However, the technique of putting diodes on the diaphragm or near the diaphragm is practical – it works. There are problems, however, and the entire system must be put into the hands of the production people.

* See chapter by A. E. Walker, L. J. Viernstein, and J. G. Chubbuck.

**Question**: I would like to understand the principle of success of this transducer. Is it the fact that we are now building a frequency-sensing device or is it really in the sensor? Could I use another sensor or another frequency-selected device?

**Dr. Fryer**: I think that both parts are essential. You have got to have an electronic system that is stable as well as a transducer that is stable. There is nothing unique about that.

**Question**: I mean the principle of reading frequency, is that a part of the success?

**Dr. Sandler**: The principle of success is that we do not need to glue the strain gauges on the diaphragm.

**Dr. Fryer**: I do not think that is his question.

**Question**: Do you mean to say that you could do the same thing with an analog readout and not with a frequency readout, and that you would have the same result?

**Dr. Fryer**: No, I do not think that you are answering the question Hal. I understand the question. I think that both parts are significant and important. You have to start with a stable transducer no matter what. I do not care how good a circuit it is, if the transducer is temperature-sensitive and you do not have a compensation, or if it is affected by other things than pressure, it is not effective. Any transducer that you design, whether it is a thermistor or whatever, you want it sensitive to only one variable, that is, the variable that you are interested in measuring. If it is sensitive to two or three other variables and you cannot correct for them, then it is no good. So you must have a good transducer. We are working with a unique aspect – a picofarad full scale, or a fraction of a picofarad. Most people have assumed that it is impossible to work with such small changes and have tried to use bellows and other arrangements. But with a suitable circuit, such small capacity changes can be accurately measured. The DC bridge circuit, with the diode rectification right at the device is what makes the system feasible for small-size catheters. When we did this 10 years ago, the diode circuit was not readily available.

Now the larger capacity cell will give more capacity and is specifically being used with the circuit and for the intracranial pressure application. It is readily amenable to a telemetry system so that you can radiate it to a distant point and lose nothing in accuracy. You do not need bridge balancing and you do not need a second circuit, subcarrier, digital pulse code modulation (PCM), or some kind of coding in order to transmit without losing accuracy – that is what I was trying to point out. The readout is on a counter and therefore is not adjustable. While with a resistive bridge-type circuit, one is always uncertain as to what balance changes may have been made.

## REFERENCE


1. **Harrison, D. R. and Dimeff, J.,** A diode-quad bridge circuit for use with capacitance transducers, *Rev. Sci. Instrum.*, 44(10), 1468, 1973.

# PRESSURE TRANSDUCER EVALUATION*

**Paul S. Lederer**

In order to assure meaningful measurements with pressure transducers, their performance characteristics must be evaluated. Static and dynamic calibration techniques, as well as environmental and durability test procedures are discussed. Reference is made to on-going work at the National Bureau of Standards, as well as to the standardization activities of professional societies.

Enough evidence has accumulated to indicate that transducers and biomedical measuring systems have certain performance limitations. In addition, there is really not enough knowledge about their performance in many instances. This is particularly true for pressure transducers.

In order to rely on the measurements made by the aid of such transducers, one must know a great deal about the performance characteristics. These performance characteristics must be found by the evaluation of transducers.

The area of evaluation can be divided into four major subdivisions:

1. Static calibration. This is the measurement base – the foundation which is used to establish the operation of the device (unless it is a piezoelectric device, in which case one has to use a dynamic calibration).
2. Dynamic calibration of the transducer. This is necessary because these devices rarely ever sense anything but dynamic phenomena, therefore, one must know how the pressure transducer responds to nonstatic pressures.
3. Investigation of the environmental effects. The environment in which the transducer operates during measurement may have a significant effect on the capability of the transducer to perform that measurement.
4. Durability. It is something one would like to know: how *long* the transducer continues to perform in the way it is supposed to perform. Certainly, implantation is one of the conditions that will have an effect on the durability and quality of measurement.

What can be derived from these different evaluation areas? In the static calibration using laboratory standards, one determines such things as linearity, hysteresis, and repeatability. One can extend the testing further by obtaining characteristics like the effects of power supply voltage on the sensitivity or on the zero output. One also could get information on warm-up drift – how soon to turn on the instrument before a measurement can be made with it.

A thorough discussion of dynamic calibration is included in the chapter by L. Geddes. Basically, one is looking for three things: amplitude-frequency response, phase-frequency response, and the dynamic linearity. Ideally, one would like to get this by way of a sinusoidal calibration, just as is done in obtaining the characteristics of an amplifier. In the case of many measurements this is not possible or practical, and one may have to resort to a different kind of dynamic input. A case in point would be the step function calibration of a pressure transducer, from which one can then derive (in various ways) a system function, which will contain the necessary phase and frequency information.

Some researchers may want to go to something more complicated. Many pressure transducers are used at the end of catheters, which are long nonrigid tubes. There will be effects on frequency response due to the mass of liquid in the tube, and there will be damping effects as well. Furthermore, these effects are also subject to temperature variations. So, it may not be enough to run a dynamic calibration at laboratory ambient conditions. One should do it at the temperature at which the actual measurement is going to be made.

As far as the environment itself is concerned, temperature is an important parameter, as are vibration and shock. However, there are also some more subtle things. For example, in regard to temperature one normally thinks of letting the transducer stabilize at a given temperature, carrying out the evaluation, then raising it to another temperature, and running another evaluation. But there are other thermal effects which can have a considerable influence on the output of the transducer, such as the situation when one takes a transducer which is at laboratory ambient conditions and sticks it into an animal, which is now at a different ambient condition, subjecting the transducer to rapid change in environment. Humidity has effects on semiconductor transducers, partly because of the fact that it may generate corrosion inside the transducer. There are also other things to consider, e.g., the fact that when one energizes a transducer, one raises the temperature a little above ambient, and that will tend to change the temperature compensation and therefore the transducer output.

Finally, in the area of durability, one would like to have something that evaluates the life of the transducer, specifically the length of time that one can rely on its measurements. Here again there are two major areas of concern. One of these areas is storage environment; most transducers will show some kind of drift, even if they just sit on the shelf. Therefore, one periodically has to recalibrate them. There is another environment which is probably more characteristic of implantation, especially in the cardiovascular system. Thus, one subjects the transducer to cycling pressures over many, many thousands of cycles, and this may have both a short-term and a long-term effect on the transducer characteristics. There are probably many other parameters which will affect the transducers' ability to perform good measurements over long periods of time.

It is important to realize that in order to be really able to rely on measurements made by a transducer, one must know something about the performance characteristics. Thus, the transducer must be evaluated adequately.

Transducer evaluation is a complex, time-consuming, and very expensive process. Yet if it is lacking, meaningful measurements with transducers are not really possible. The cost of the transducers is only a fraction of the cost of the measurements that are made with them. It would be wrong to simply conclude that the cost-effectiveness of any given approach is related only to the cost of the transducer. Other factors should be considered. A *system cost* is involved and the researcher's own evaluation or periodic recheck is a part of it, as well as phone calls to the manufacturers to get help, and looking into data trying to analyze exactly what they mean and how they relate to what is really being measured.

The National Bureau of Standards plays a significant role in transducer evaluation. NBS is able to calibrate secondary standards or instruments in conformance with the national standards. This is done in a variety of areas of measurement, although there is relatively little dynamic calibration going on except in such areas as vibration pickups and microphones.

Transducer evaluation started at NBS in 1951 when, at the request of the Bureau of Aeronautics of the Navy, a small project for the evaluation of transducers and the development of evaluation methods was initiated. The Navy was interested in the performance of telemetry transducers which were used in test aircraft at that time. This work has been going on since 1951 and has developed a number of evaluation and calibration techniques and prototype equipment. A number of publications have resulted from this work.[1-20] We have evaluated a small number of transducers as a by-product of our main output – the development of evaluation methods. Ultimately, the results of the evaluations and the development of the evaluation methods have been used in the development of standards, primarily those of the Instrument Society of America, and the American National Standards Institute. Information has also been disseminated in workshops sponsored by the NBS. Eight transducer workshops have been held (the last one in Dayton, Ohio early in 1976); however, they were all primarily aerospace-oriented and did not include information on biomedical instrumentation. Some of the equipment developed as part of the NBS transducer project is shown in Figures 1 through 6. Bear in mind that this project was directed to aerospace needs and applications.

Figure 1 demonstrates something fairly recent: a dynamic pressure calibrator using a liquid medium. It basically consists of a tube filled with a liquid which is vibrated in the vertical plane. If one knows the density of the liquid, the height of the liquid above the transducer, and the acceleration applied to the whole system, one can calculate the applied pressure. This particular system has been used for pressure levels of the order of 7 to 10 psi. With a 9-in. tube a natural frequency of about 1,600 Hz results, which means a flat response to about 40% of that. Using different liquids and different heights, one can get different natural frequencies. There are two limitations with this method: (1) no matter how good a liquid is used, with a high velocity of the propagation of sound, the ultimate limitation is the deflection of the diaphragm of the transducer, which will tend to drag down the system frequency; and (2), the

FIGURE 1. Liquid medium dynamic calibrator for pressure transducers. (From Vezzetti, C. F., Hilten, J. S., Mayo-Wells, J. F., and Lederer, P. S., A New Dynamic Pressure Source for the Calibration of Pressure Transducers, NBS Technical Note 914, National Bureau of Standards, Washington, D.C., June 1976.)

transducer is vibrated, although it is mounted in such a way that the vibration is across the diaphragm, which is the least sensitive direction for vibration response of the transducer. Nevertheless, if one uses this kind of a system, one must look at the vibration output of the transducer before the system is filled with liquid.

Figure 1 shows a large electromagnetic shaker with the vertical tube mounted on the shaker. Since most transducer laboratories have a shaker system, the addition of this tube structure is a relatively simple thing and gives them dynamic pressure calibration capability using a liquid medium.

Figure 2 shows a step function pressure calibrator using a quick-opening valve. This particular device was built for the Naval Air Test Center and generates a step function of pressure (actually a ramp) with a rise time of about 1.5 ms. The pressure generated can be anywhere from 1 to 100 psi. The long tube coming out of the front end was added at the request of the Navy so that they can take this calibrator out to the flight line and calibrate a transducer built into a wing by simply holding the end-fitting against the transducer. When the pressure line is added, the rise time of the calibrator degrades to approximately 15 to 20 ms.

FIGURE 2. Quick-opening valve dynamic calibrator for pressure transducers. (From Lederer, P. S., Step Function Pressure Calibrator, U.S. Patent 3,034,332, May 15, 1962.)

Figure 3 shows a shock tube used for nanosecond pressure steps with amplitudes up to 1,000 psi.

Figure 4 illustrates a job done for the NASA Lewis Laboratory. They were interested in the durability of the group of pressure transducers. We cycled these transducers at ambient temperature, and at 150°F, for 40 million cycles and developed some very interesting data. Again, the main purpose was to develop the test method and methods for interpreting the data. It is also interesting to note that of the 20 pressure transducers we tested, not a single one failed; however, a number of stepping motors, solenoid valves, and counters were lost.

One of the problems in using pressure transducers in aerospace research is making measurements in the presence of blast pressures near jet engines. One encounters very large thermal transient inputs. In order to simulate this effect a continuous wave laser was used, whose beam impinged on the diaphragm of the transducer (Figure 5). A mechanical shutter was used to interrupt the beam, and the transducer was exposed for approximately 2 min. During this period pressure was alternately switched on and off, with the beam going through a glass window in front of the transducer diaphragm. It was possible to obtain very nice recordings showing both the zero and sensitivity change as a function of time during the illumination with the laser beam. The energy in the laser beam was measured with a radiant power meter. There was fairly good correlation between thermal power input vs. both zero and sensitivity shift. One transducer tested was a semiconductor device with strain gauges diffused into the diaphragm, and the expected change in output resulted. Once the excitation supply to this particular transducer was accidentally not turned on, resulting in a very similar output. This transducer turned out to be a better photo cell than pressure transducer.

Since the National Test Ranges requested inexpensive, rapid, thermal transient test method, NBS experimented with photographic flash bulbs and had a radiometer especially modified to measure the energy in a short-duration flash, lasting on the order of 13 to 15 ms. The setup shown in Figure 6 was developed for the testing of piezoelectric accelerometers.

To summarize, the output that the National

FIGURE 3. Shock tube for dynamic calibration of pressure transducers. (From Smith, R. O. and Lederer, P. S., A Shock Tube Facility for Dynamic Calibration of Pressure Transducers, NBS Report 5941, National Bureau of Standards, Washington, D.C., July 1958.)

Bureau of Standards provides is the *test method.* NBS describes the procedure, making it possible for others to test their own transducers using this method.

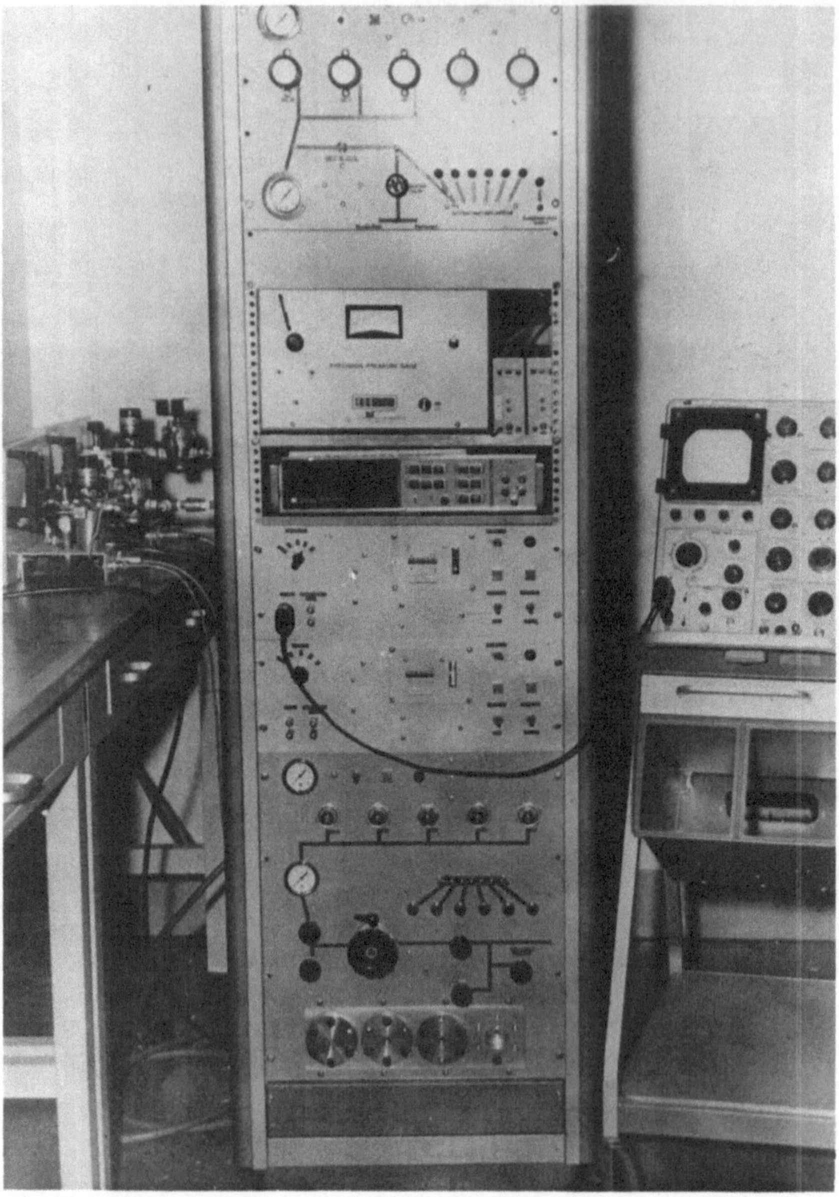

FIGURE 4. Apparatus for life-cycling tests on pressure transducers. (From Lederer, P. S. and Hilten, J. S., Durability Investigation of a Group of Strain Gage Pressure Transducers, NBS Report 10975, National Bureau of Standards, Washington, D.C., December 1972.)

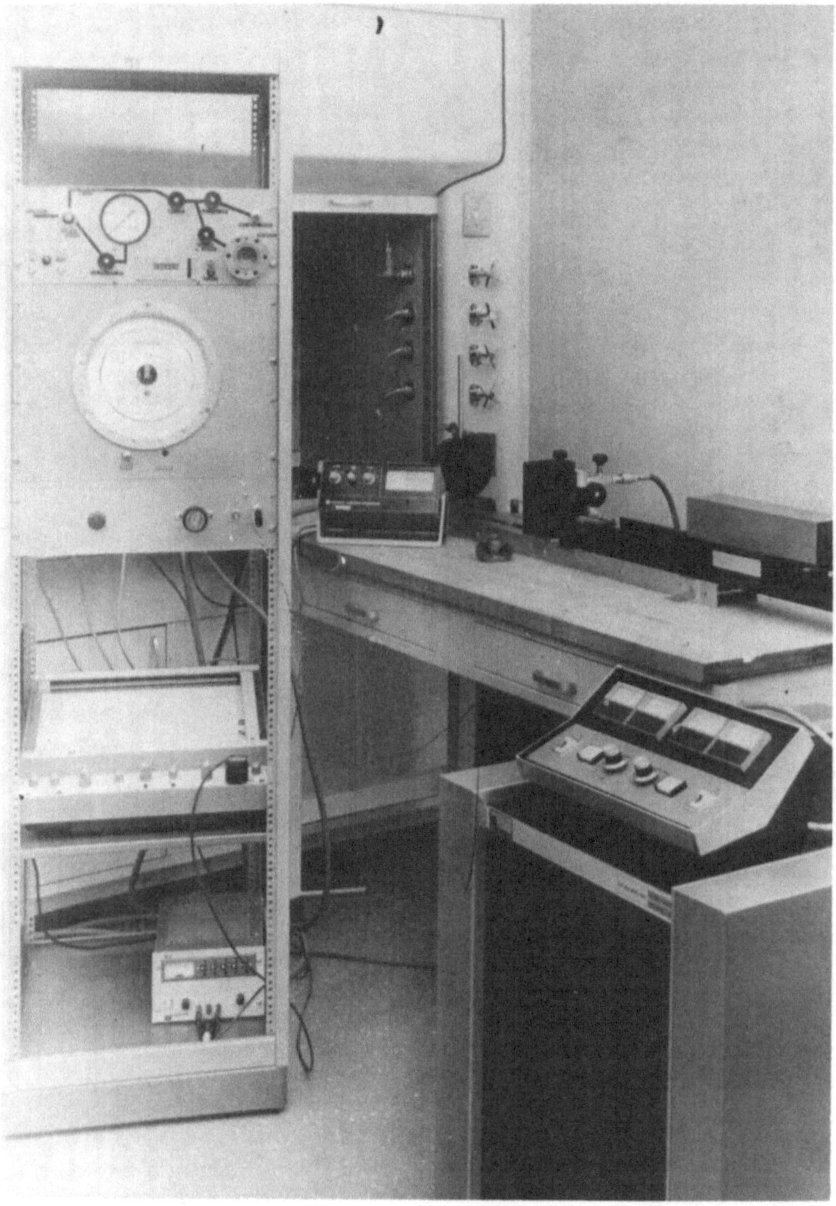

FIGURE 5. Laser setup for study of thermal transient effects on pressure transducers. (From Lederer, P. S. and Hilten, J. S., A Laser Technique for Investigating the Effects of Thermal Transients on Pressure Transducer Performance Characteristics, NBS Technical Note 723, National Bureau of Standards, Washington, D.C., May 1972.)

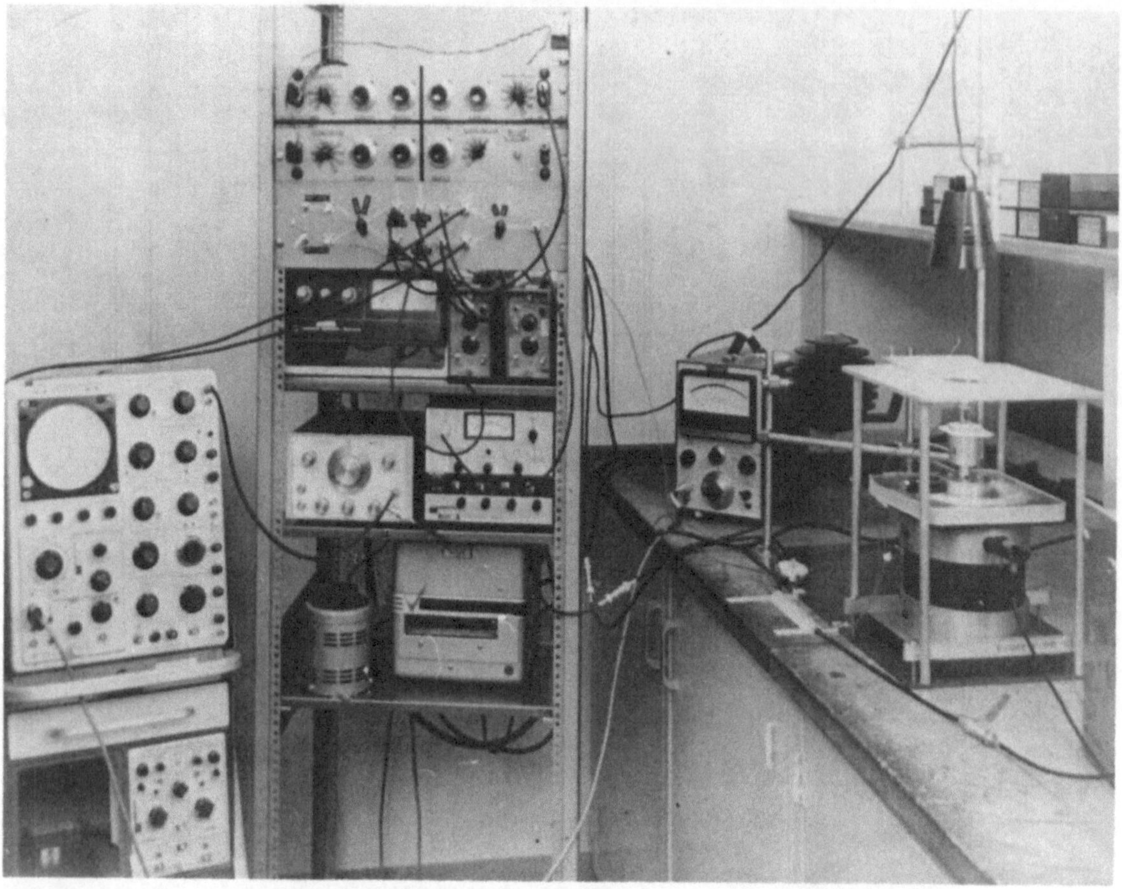

FIGURE 6. Heat lamp setup for study of thermal transient effects on piezoelectric accelerometers. (From Vezzetti, C. F. and Lederer, P. S., An Experimental Technique for the Evaluation of Thermal Transient on Piezoelectric Accelerometers, NBS Technical Note 855, National Bureau of Standards, Washington, D.C., January 1975.)

# REFERENCES


1. **Lederer, P. S.**, Step Function Pressure Calibrator, U.S. Patent #3,034,332, May 15, 1962.
2. **Lederer, P. S.**, General Characteristics of Linear Strain Gage Accelerometers Used in Telemetry, NBS Technical Note 150, National Bureau of Standards, Washington, D.C., June 1962 (out of print).
3. **Smith, R. O., Willis, E. A., and Hilten, J. S.**, A dual centrifuge for generating low-frequency sinusoidal accelerations, *J. Res. Nat. Bur. Stand.*, 66C(4), 357, 1962.
4. **Schweppe, J. L. et al.**, Methods for the Dynamic Calibration of Pressure Transducers, NBS Monograph 67, National Bureau of Standards, Washington, D.C., December 1964 (out of print).
5. **Smith, R. O.**, A liquid medium step function pressure calibrator, *J. Basic Eng. Trans. ASME*, 86D(4), 723, December 1964.
6. **Lederer, P. S. and Hilten, J. S.**, Earth's Field Static Calibrator for Accelerometers, NBS Technical Note 269, National Bureau of Standards, Washington, D.C., February 1966.
7. **Lederer, P. S.**, Methods for Performance-Testing of Electromechanical Pressure Transducers, NBS Technical Note 411, National Bureau of Standards, Washington, D.C., February 1967 (NTIS TN 411).
8. **Lederer, P. S.**, "Life Cycling" Tests on Several Strain Gage Pressure Transducers, NBS Technical Note 434, National Bureau of Standards, Washington, D.C., October 1967 (NTIS TN 434). Lederer, P. S., NBS Inter Agency Transducer Project, NBS Technical Note 469, National Bureau of Standards, Washington, D.C., October 1968.
9. **Lederer, P. S.**, Pressure – introduction and state-of-the-art, in *ISA Transducer Compendium*, 2nd ed. part 1, Harvey, G. F., Ed., Plenum, New York, 1969.
10. **Newman, N.**, Considerations in Computing the Useful Frequency Range of Piezoelectric Accelerometers, NBS Technical Note 487, National Bureau of Standards, Washington, D.C., July 1969 (NTIS TN 487).
11. **Horn, L.**, Thermal Gradient Effects on Thirteen Flush Mounted Pressure Transducers, NBS Technical Note 490, National Bureau of Standards, Washington, D.C., August 1969 (NTIS TN 490).
12. **Williams, R.**, The Effects of Extended High-temperature Storage on the Performance Characteristics of Several Strain Gage Pressure Transducers, NBS Technical Note 497, National Bureau of Standards, Washington, D.C., October 1969 (NTIS AD 695821).
13. **Hilten, J. S.**, Accelerometer Calibration with the Earth's Field Dynamic Calibrator, NBS Technical Note 517, National Bureau of Standards, Washington, D.C., March 1970 (NTIS PB 190072).
14. **Hilten, J. S., Lederer, P. S., and Sethian, J. D.**, A Simple Hydraulic Sinusoidal Pressure Calibrator, NBS Technical Note 720, National Bureau of Standards, Washington, D.C., April 1972 (GPO $0.65).
15. **Lederer, P. S. and Hilten, J. S.**, A Laser Technique for Investigating the Effects of Thermal Transients on Pressure Transducer Performance Characteristics, NBS Technical Note 723, National Bureau of Standards, Washington, D.C., May 1972 (GPO $0.85).
16. **Lederer, P. S. and Hilten, J. S.**, Durability Investigation of a Group of Strain Gage Pressure Transducers, NASA CR 120953 NBS Report 10975, Washington, D.C., December 1972.
17. **Lederer, P. S.**, Transducers in the Real World, Joint Measurement Conference, National Bureau of Standards, Gaithersburg, Maryland, November 1974.
18. **Vezzetti, C. F. and Lederer, P. S.**, An Experimental Technique for the Evaluation of Thermal Transient Effects on Piezoelectric Accelerometers, NBS Technical Note 855, National Bureau of Standards, Washington, D.C., January 1975 (GPO $1.05).
19. **Hilten, J. S., Vezzetti, C. F., Mayo-Wells, J. F., and Lederer, P. S.**, A Test Method for Determining the Effect of Thermal Transients on Pressure-Transducer Response, NBS Technical Note 905, National Bureau of Standards, Washington, D.C., April 1976.
20. **Vezzetti, C. F., Lederer, P. S., and Hilten, J. S.**, A New Dynamic Source for the Calibration of Pressure Transducers, NBS Technical Note 914, National Bureau of Standards, Washington, D.C., May 1976.

# REFERENCES


[illegible]

# *Future Directions for Implantable Pressure Transducers*

# FUTURE DIRECTIONS FOR IMPLANTABLE PRESSURE TRANSDUCERS: THE USER'S POINT OF VIEW

**Harold Sandler, M.D.**

The measurement of pressure for physiological assessment of function has now become increasingly important. Studies utilizing chronically instrumented animals and long-term observation of body function are becoming more commonplace. Evaluation of disease processes or drug actions presently provides the strongest stimulus for such information. User needs have been for transducers that are increasingly smaller, work reliably and accurately, and cost less in order to allow their more frequent and widespread use. Technological developments demonstrate that such goals are feasible. Present systems are being used as chronically implantable systems in animals. With continued support and progress of these activities, these devices can and will eventually be made available for use in man.

When discussing the problems of measuring pressure (force per unit area), many aspects must be considered. The system is not simple and relies on factors other than simply the transducer – there are needs for attention to the systems used in processing the signal and means for their recording or display as well. Obtaining pressure records of the highest quality demands time and care. With technological progress, the demand by clinician and physiologist has been to maintain standards, but relieve tedium. The user has asked that the transducer and its associated physiological monitoring systems become smaller and more reliable; work accurately for long periods of time; and constantly cost less, therefore becoming more readily available. These goals are often mutually exclusive. Despite these problems, significant progress has been made. The significant driver to this end has been the importance of making physiologic pressure measurements. It is, therefore, the purpose of this chapter to briefly summarize how pressure measurements are being used and to discuss future needs.

Medical use is characterized by a diversity of needs for measurement of force, whether under clinical or experimental conditions. Every organ system within the body has been studied in this regard; the most frequently studied are the cardiovascular, gastrointestinal, genitourinary, and reproductive systems. Resultant measurements have been used to describe function as shown in Figure 1. Elastic or mechanical properties of tissues have been described by determining pressure-volume or stress-strain relationships. This has been done specifically for solid organs where size and shape could be related to measured intracavitary pressure and/or measured or calculated intramural forces.[1–3] There are also other considerations for several organ systems where static loads may critically determine function, such as cardiac preload,[4] cerebrospinal fluid pressure, and intraocular pressure.[5] Last, but not least, it is becoming increasingly clear that differential pressure across the walls of hollow organs is an important functional variable and needs to be detected throughout the respiratory cycle and at various sites along the cardiovascular system.[6–9] All hollow organs also contain transmural pressure gradients across their walls, whose alterations may result in significant functional or adaptive changes.[6,9,10] In-depth consideration of these uses is beyond the scope of this discussion. The

MECHANICAL PROPERTIES OF TISSUE
- PRESSURE – VOLUME
- STRESS – STRAIN

STATIC LOADS
- CARDIAC PRELOAD
- CSF
- INTRAOCULAR
- INTRAMURAL

DIFFERENTIAL
- RESPIRATORY EFFECTS
- HYDRAULIC FLOW
- TRANSMURAL

FIGURE 1. Need for making pressure measurements.

area where pressure measurements have received their largest and widest use has been in determining cardiovascular function. The remainder of the material in this chapter will, therefore, be limited to a discussion of such applications. This in no way negates application to other body areas. With only rare exception, systems developed for vascular usage can be readily adopted to the other organ systems just described.

The requirement for frequency response in order to accurately reproduce wave forms during intravascular pressure measurement is now well established.[11] A well-designed system must be capable of flat response to 20 Hz or the tenth harmonic at heart rates up to 120 beats/min (2 Hz). The effect of omitting harmonics higher than the tenth is small (1 to 2%). At lower heart rate levels, the bandwidth or frequency range will be less. Much has been accomplished with just such measurements of arterial or ventricular pressures. By applying the laws of Pascal and Poiseuille, the relationship between pressure and flow has been established, and subsequently the ability to treat pulsatile flow in precise mathematical terms.

Cardiac pump function has required determination of preload (filling pressure for the ventricles),[4] peak ventricular pressure, rate of change in pressure (dp/dt),[12] and afterload or impedance faced by the ejecting ventricle.[13] The importance of filling pressure in determining subsequent ventricular performance is embodied in Starling's Law of the Heart – a change in stretch for cardiac muscle, evidenced by increased filling pressure or volume, results in increased force and/or stroke volume. Ventricular filling pressures range normally between 3 to 5 mmHg (range 3 to 12 mmHg) and are venous in nature. These pressures have proven to be the most difficult to accurately measure at rest, particularly in the intact animal and man. The problem has become more complicated with the need to detect changes during pharmacologic or exercise-induced stresses. Motion artifact in such latter cases has proven to be so great in fluid-filled systems as to often destroy the accuracy of such measurements. Respiratory effects have also produced problems, particularly if it is realized that transmural pressure changes may be the important physiologic variable needed for measurement in these instances. Use of catheter-tipped transducers has also been hampered by their sensitivity to blood temperature changes and to base-line instability caused by alterations in blood flow. It is clear that for animal studies, an implantable, stable, accurate right atrial or left ventricular pressure cell is the approach of choice. However, the accuracy required is to an absolute level of 0.1 mmHg, and base-line drift over the course of a study would be intolerable. Stability for longer periods of weeks and days is required if long-term studies are to be accomplished without need for repeated heart catheterization. A distinct disadvantage with present implantable systems is their design to operate as absolute pressure transducers. In this regard, changes in base line may not represent electrical instability, but rather normal changes in barometric pressure occurring with climate change. As previously reported from our laboratory[14] this can serve as a distinct advantage in certain cases, since dynamic gain for the transducer system can be checked in a totally implanted system by sensing response to changes in barometric pressure in a homemade or commercial altitude chamber.

If an abrupt change in arterial pressure is made during a period when preload and contractility are constant, the stroke output of the heart changes inversely as a function of the average impedance to ejection. Similarly, if the impedance to ejection is increased, stroke volume can be maintained only by an increased filling pressure. Again, the need for pressure measurement to determine function is clear. Cardiac power can be computed when such flow and pressure values are known, and it is computed as their cross-product. Despite near ideal performance of modern pressure transducers, the use of hydraulically coupled systems to obtain pressure under these circumstances is fundamentally limited in frequency responses by the compressibility of the fluid used. Therefore any insoluble gas bubbles trapped in the fluid will have serious effects in distorting resultant signals. These systems also suffer from acceleration artifacts associated with the rapid movement of the catheter. Although such systems can be designed to cover the frequency ranges associated with blood pressure waveforms, it is difficult to do this for the frequency range associated with cardiovascular sounds. Although intracardiac phonocardiography has been previously explored to detect alterations in valve function in congenital or rheumatic heart disease, its most useful future purpose may lie in assessing cardiac function.[15] It has been realized by several investigators that the character of intracardiac sounds (particularly the first heart sound) may be altered with coronary artery disease.[16] It has also been shown that altered cardiac material wall properties can result in altered ability to generate and propagate sound. Implanted pressure cells or manometer-tipped catheters capable of wide band response (0.1 to 4 kHz) with appropriate band-pass filtering may serve to provide the dual function of simultaneous sound (100 Hz to 4 kHz) and pressure (0.1 Hz to 100 Hz) detection.

In regard to implantable pressure transducers, user requirements are shown in Figure 2. Size will continue to be a problem as smaller animals are used. The need for transducers 1 mm in diameter will increase and become mandatory. With small size comes the requirement for minimal weight and body reaction when working in small rodents and non-human primates. Implantable transducers for these uses will have high initial costs. They must also be capable of long life and/or repeated use without repair. Useful transducer function for periods of 6 months to a year after once being implanted seems reasonable and necessary for most long-term animal experiments where progress of disease, such as atherosclerosis or physiologic state during space flight, will be studied. With smaller size must come the eventual design requirements of battery operation – first from transmitter units outside the body (backpack) and eventually from battery power implanted at the time of surgery. For those applications which will span periods greater than 6 months to a year, it will be beneficial to expect designs that do not require batteries at all, but depend on radio frequency induction of required voltage.[17] Systems should ideally be disposable and of reasonable cost; this is particularly the goal if large numbers of small animals, such as rodents, are used. The units can then be considered part of the operating costs for conducting the study or procedure. Ability to sense and faithfully reproduce filling (venous) pressure as well as arterial pressure will be important in many experiments; these data will be collected for months in most animals. Finally, it is becoming increasingly clear that transmural or differential pressure is important. In these cases, the capability to sense pericardial and/or intrathoracic pressures and to compare these instantaneously with atrial and/or ventricular values to determine vascular or mechanical function will be necessary.

MINIMAL SIZE
MINIMAL WEIGHT
MINIMUM BODY REACTION – NO CLOTS
LONG LIFE
LOW POWER – BATTERY LESS IF POSSIBLE
RELIABLE – BASELINE STABILITY
CAPABLE OF USE WITH TELEMETRY LINK
SENSE VENOUS AS WELL AS ARTERIAL
DISPOSABLE
SENSE DIFFERENTIAL

FIGURE 2. User requirements for implantable pressure transducers.

Ready access to cardiac chambers and central circulation in man is possible only by cardiac catheter. Previous needle-puncture studies of cardiac chambers through the chest wall have carried significant risk and have proven to be of limited value. The need for higher fidelity pressure recordings and the potential for compound recordings, such as the electrocardiogram, flow, and/or heart sounds at selected sites, have promoted a shift towards manometer-tipped catheters. It is of interest to note that the first electrical blood pressure transducer was of the catheter-tip type; it was developed by Grunbaum in 1898.[18] Over the past 10 years a large number of devices using various transducer principles on the ends of catheters have been tested and used in animals and humans (see Figure 3). The sensing principle is listed on the left-hand side of the figure; the transducer type is on the right. The Telco® catheter was one of the first commercially available compound catheters. In addition to pressure, ECG and phonocardiograms could be sensed as well. Both the Statham® and Millar catheters contained lumens for calibration using a fluid column; the Konigsberg device contains no means for external calibration. The capacitance, optical, and solid state catheters are all capable of use as differential pressure-measuring systems. The capacitance, optical, and tunnel diode catheters have been made to a size of 1 mm in diameter; silicon diaphragm devices are capable of smaller size (0.5 to 1.0 mm). Tunnel diode transducers have had the problem of temperature instability.

In each case, the objective for development was to allow the user to obtain an increasingly accurate waveform with greatest frequency response. Much of this was prompted by the need to obtain the rate of change of ventricular or aortic pressure waveforms. Here a need for 25 Hz frequency response has been established as well as use of at least eight harmonics to detect to a 5% level of accuracy.[11,12] These conditions are not readily available even with the best fluid-filled systems.

| | | |
|---|---|---|
| INDUCTANCE | – | TELCO – ECG, PHONO |
| RESISTANCE | – | STATHAM, MILLAR, KONIGSBERG |
| CAPACITANCE | – | DIFFERENTIAL PRESSURE |
| OPTICAL | – | DIFFERENTIAL PRESSURE |
| SOLID STATE | – | SILICON DIAPHRAGM<br>TUNNEL DIODE |

FIGURE 3. Varieties of manometer-tipped catheters.

| | | |
|---|---|---|
| INDUCTANCE | – | McKAY, COLLINS |
| RESISTANCE | – | KONIGSBERG |
| PIEZORESISTIVE | – | KO |
| N-P-N | – | PITRAN |
| BELLOWS | | |
| CAPACITANCE | | |

FIGURE 4. Implantable transducers.

These needs became stronger as more accurate definitions of cardiac performance were being made, and there was a shift from concern with hydraulic cardiac performance to concern for the characteristics of intact cardiac muscle and its mechanical properties. The previous hydraulic system concerned with the parameters of pressure, volume, and flow was described in terms of force, length, and velocity. The interrelationships between these latter variables and the index used to judge ventricular performance were termed "contractility," and dp/dt has been used to determine this state.

The present choices of pressure transducers available for implantation use are listed in Figure 4. The earliest relied on transduction by measurement of inductance and were used by McKay[19] for swallowable capsules and later implanted to sense arterial pressure. Collins later used this principle for measuring intraocular pressure.[5] At present, the transducer most commonly used commercially is manufactured by Konigsberg Instruments.

Now it is appropriate to consider the future direction of pressure measurement. Projections for its use in humans and animals are listed in Figures 5 and 6 respectively.

For animal work there is a need for continued development and use of implantable systems. The transducers must be continually decreased in size from the now commercially available sizes (2.5 mm to 6.5 mm) to 0.5 to 1.0 mm in diameter. This will allow their use in small rodents and

DECREASE SIZE – USE MICE, RATS
LONG LIFE – 1 - 2 years
BATTERY LESS OPERATION
DISPOSABLE
LOW PRESSURE ACCURACY
MULTIPLE INTRAMURAL
MULTIPLE INTRAVASCULAR

FIGURE 5. Future directions of animal transducers.

IMPROVED CATHETER TIPPED TRANSDUCERS
DECREASE SIZE – 0.5 to 1 mm diameter
BATTERY OPERATION
DISPOSABLE
DIFFERENTIAL
CALIBRATABLE IN VIVO
IMPLANTABLE

FIGURE 6. Future directions for human use.

primates – in the former case for experiments needing large numbers of observations; and in the latter for processes simulating man. Stability and accuracy need attention and improvement in both cases. A decrease in cost would be necessary if such widespread use is contemplated; disposability of the entire system would be the ideal goal and would allow for multiple transducer use in a single animal. Uses in man will benefit from advances made in animal capability, first to create smaller, more reliable manometer-tipped catheters and later for use as implants left at the time of surgery. Thus, catheters should be small in size and have the ability to measure multiple parameters from a single system (compound catheter). The ability to band-pass a high-fidelity pressure signal to obtain simultaneous phonocardiography and pressure is well known. Equally well demonstrated has been the technique for flow determination, using information from the change in pressure measured from sensors at a known fixed distance apart. The objective is to miniaturize these sensors so that they may be used on a catheter introduced into a coronary vessel in animals or humans. Lastly, it is projected that with the advent of small, reliable, batteryless units, transducers will be designed that may be left in humans after surgery to detect pressure for the remainder of the patient's life. These can take two forms: the first will be a cuff placed around selected vessels, using a transducer external to the wall; the second will be a cuff-link shaped system whose base will contain the transmission and receiving antennae, and whose shaft will protrude into the blood stream.

The future appears bright. The user community wants a reliable, cheap transducer which can be implanted or used as a catheter. Technological developments presented by several investigators demonstrate that such goals are now feasible. With continued support of these activities, these devices can, and will, be made more generally available and eventually commercialized.

## REFERENCES


1. **Sandler, H. and Dodge, H. T.**, Left ventricular tension and stress in man, *Circ. Res.*, 13, 91, 1963.
2. **Villanueva, C. and Sauvage, J. P.**, Intrauterine pressure monitoring with a ballon tipped catheter, *Obstet. Gynecol.*, 45, 287, 1975
3. **Lewartowski, B., Sedeh, G., and Okolska, A.**, Direct measurement of tension within the left ventricular wall of the dog heart, *Cardiovasc. Res.*, 6, 28, 1972.
4. **Sarnoff, S. J.**, Myocardial contractility as described by ventricular function curves; observation on Starling's Law of the Heart, *Physiol. Rev.*, 35, 107, 1955.
5. **Collins, G. C.**, Miniature passive pressure transducer for implanting in the eye, *IEEE Trans. Biomed. Eng.*, BME-14, 77, 1967.
6. **Glantz, S. and Kernoff, R.**, Muscle stiffness can be determined from diastolic pressure-volume curves, *Circ. Res.*, 37, 787, 1975.
7. **Avasthey, P., Coulam, C. M., and Wood, E. H.**, Position dependent regional differences in pericardial pressures, *J. Appl. Physiol.*, 28, 622, 1970.
8. **Porjé, I. G. and Rudewald, H.**, Hemodynamic studies with differential pressure techniques, *Acta Physiol. Scand.*, 51, 116, 1961.
9. **Baird, R. J., Goldback, M. M., and De laRoicha, A.**, Intramyocardial pressure. The persistence of its transmural gradient on the empty heart and its relationship to myocardial oxygen consumption, *J. Thoracic Cardiovasc. Surg.*, 64, 635, 1972.
10. **Coulam, C. M. and Wood, E. H.**, Regional differences in pleural and esophageal pressure in head-up and head-down positions, *J. Appl. Physiol.*, 31, 277, 1971.
11. **Gabe, I. T.**, Pressure measurement in experimental physiology, in *Cardiovascular Fluid Dynamics.* Vol. 1. Bergel, D. H., Ed., Academic Press, New York, 1972, 11.
12. **Gersh, B. J., Hahn, C. E. W., and Prys-Roberts, C.**, Physical criteria for measurement of left ventricular pressure and its first derivative, *Cardiovasc. Res.*, 5, 32, 1971.
13. **Milnor, W. R.**, Arterial impedance as ventricular afterload, *Circ. Res.*, 36, 565, 1975.
14. **Sandler, H., Stone, H. L., Fryer, T. B., and Westbrook, R. M.**, Use of implantable telemetry systems for study of cardiovascular phenomena, *Circ. Res.*, 31, 11, 1972.
15. **Piemme, T. E., Barnett, G. O., and Dexter, L.**, Relationship of heart sounds to acceleration of blood flow, *Circ. Res.*, 18, 303, 1966.
16. **Price, W. H. and Brown, A. E.**, Alteration in intensity of heart sounds after myocardial infarction, *Br. Heart J.*, 30, 835, 1968.
17. **Fryer, T. B. and Sandler, H.**, A rechargeable battery system for implanted telemetry systems, *Proc. Conf. Eng. Med. Biol.*, 12, 129, 1970.
18. **Grunbaum, O. F.**, On a new method of recording alterations in pressure, *J. Physiol.* (London), 22, 44, 1897-8.
19. **McKay, R. S.**, Telemetering from within the body of animals and man: Endoradiosondes in biomedical telemetry, in *Biomedical Telemetry,* Caceres, C. A., Ed., Academic Press, New York, 1965, 148.

# REFERENCES

# *Indices*

# AUTHOR INDEX

## G



## H



## I



## J



## K



## L



## M

## N



## O



## P



## R



## S



## T

## U



## V



## W



## Y



## Z

# INDEX

E



F



G



H



I

## K



## L



## M



## N



## O



## P



## R



## S

## T



## U



## V



## Z